C. M. White

Modern Supercritical Fluid Chromatography

Chromatographic Methods

Series editors

W. Bertsch

W. G. Jennings

Modern Supercritical Fluid Chromatography

Edited by C. M. White

With contributions by

M. Ashraf-Khorassani, D. S. Boyer, S. J. DeLuca, R. B. Denyszyn, P. R. Griffiths, J. P. Guzowski, H. H. Hill, Jr., R. K. Houck, R. P. Khosah, E. Klesper, J. M. Levy, K. L. Maguire, M. Morrissey, F. Pacholec, G. L. Pariente, A. C. Rosselli, F. P. Schmitz, H. E. Schwartz, R. D. Smith, L. T. Taylor, K. J. Voorhees, H. M. Widmer, B. W. Wright, and S. D. Zaugg

SFC-87 was supported by

SOCIETY FOR ANALYTICAL CHEMISTS OF PITTSBURGH
SACP

Dr. Alfred Hüthig Verlag
Heidelberg Basel New York

Volume Editor
C. M. White
Department of Energy
Pittsburg Energy Technology Center
P.O. Box 10940
Pittsburgh, Pennsylvania 15236

CIP-Titelaufnahme der Deutschen Bibliothek

Modern supercritical fluid chromatography / ed. by C. M. White. With contributions by M. Ashraf-Khorassani . . . — Heidelberg ; Basel ; New York : Hüthig, 1988
(Chromatographic methods)
ISBN 3-7785-1569-1
NE: White, Curt M. [Hrsg.]; Ashraf-Khorassani, M. [Mitverf.]

Preface

Supercritical fluid chromatography (SFC) was first proposed in 1958 by *James Lovelock* while at Yale University in Connecticut. He conceived the idea of using supercritical fluids as chromatographic mobile phases in open-tubular columns to increase the solvent power of the mobile phase in order to permit the elution of nonvolatile substances. He related his thoughts concerning SFC to his coworkers, *Sandy Lipsky* and *Ray Landowne*, who urged him to commit them to paper. He did, and had the document notarized and witnessed. The document is included herein as **Figure 1** of the Preface. It should be mentioned that *James Lovelock* did not attempt to practice SFC, and, to my knowledge, did not publish his ideas concerning SFC. *Ernst Klesper*, an author of Chapter 1 of this book, was the first person to use a supercritical fluid as a chromatographic mobile phase with the express purpose of utilizing it for the analysis of nonvolatile materials. However, chromatographers were studying the use of high-pressure and supercritical mobile phases before *Klesper*. In particular, the work of *Riki Kobayashi* at Rice University can be cited. *Riki Kobayashi* was investigating high-pressure equilibria by chromatography, but was not using the increased solvent power of the media to elute nonvolatile materials.

Even though SFC has a long history, it remained a relatively underdeveloped area of chromatography until the 1980's when commercial instrumentation became generally available, and open-tubular SFC began to be widely practiced (primarily through the efforts of *Milton Lee* and *Milos Novotny*). Thus, SFC is still in its infancy and is now beginning a period of rapid growth. This is a particularly exciting time that presents the opportunity to perform ground-breaking research in chromatography, an opportunity not as generally available in the more mature forms of chromatography. With respect to SFC, these are times of great opportunity, of rapid growth, of new developments, and of controversy. SFC-87 and this book chronicle these exciting times. In some cases, SFC-87 and this book have been an opportunity to both set the record straight, and to set the record.

This book represents an attempt to collect some of the papers presented at SFC-87 into a concise volume that documents the state-of-the-art in 1987 and provides a foundation for future developments and future meetings. The book contains chapters by *Ernst Klesper* (the father of SFC) and his coworker *Franz Schmitz*. *Frank Pacholec* and his coworker from Suprex have prepared an excellent chapter on packed column technology. *Karen Maguire* and *Bob Denyszyn* from Scott Specialty Gases have prepared an in depth contribution on the solubility of organic modifiers in carbon dioxide. These chapters are followed by three chapters concern-

CRITICAL STATE CHROMATOGRAPHY

Gas chromatography has a performance in terms of speed of analysis and resolution far exceeding that of liquid chromatography; its scope, however, is restricted to those compounds with an appreciable vapour pressure at temperatures up to 300°C. Higher temperatures are of course theoretically possible but few substances or stationary phases are stable at the higher temperatures. Ionic compounds for example, salts and amino acids which have negligible vapour pressures are not therefore susceptible to analysis by gas chromatography.

Ionic compounds are soluble in certain gases in their critical state. Solvents such as water, ammonia, sulphur dioxide, hydrogen fluoride and others, which possess dielectric constants high enough to dissolve ionic substances in the liquid state also are able to dissolve ionic substances in the critical state. By using one of these solvents in its critical state as the moving phase the scope of gas chromatography can be extended to include ionic compounds which are otherwise involatile. The temperatures and pressures required for critical state chromatography are in the region 100° to 400°C and 100 to 500 atmospheres pressure. The use of capillary tube columns, consisting of a narrow bore metal tuve with a layer of stationary phase on its inner surface, however, is a simple matter. A simple detector for critical state chromatography is any device able to measure the change in conductivity of the gas as the ionic components emerge.

State of Connecticut
County of New Haven
Mary G. DeCroze
October 23, 1958

October 23rd 1958
James E Lovelock
Witnessed: S.R. Lipsky
R.A. Landowne

Figure 1

ed with detector technology written by *Kent Voorhees*, *Steve Zaugg*, and *Stephan Deluca* on SFC-MS, a chapter by *Gil Pariente* and *Peter Griffiths* on SFC-FTIR, and one by *Herb Hill* and *Mark Morrissey* on the use of ion mobility spectrometry as an SFC detector. These chapters are followed by those concerned primarily with the applications of SFC including those by *M. Ashraf-Khorassani* and *Larry Taylor* who use SFC for the separation of nitrogen bases using a variety of stationary phases, and an outstanding chapter produced by *Herb Schwartz* from Brownlee Labs and *Joe Levy* and *J. Guzowski* of BP America concerning the applications of SFC in the petroleum industry. A contribution by *Robin Khosah* from Alcoa on separations with supercritical fluids and alumina is included. Lastly, *H. Michael Widmer* has written an outstanding chapter on the use of SFC as a tool for industrial problem solving. *Michael Widmer* is an excellent person to write this chapter because of his vast experience in both packed and open-tubular SFC. His chapter makes a direct comparison between analytical results obtained by both SFC and high-temperature GC using on-column injection for a variety of samples. This comparison clearly documents the results obtained by SFC. Careful unbiased study of *Widmer's* results, even by SFC's most virulent critics, will show that in some cases presented herein, SFC produces better results than high-temperature GC, even when on-column injection has been used.

SFC-87 was held in Pittsburgh, PA at the Hyatt Hotel on Oct. 1 and 2, 1987. SFC-87 was organized by the Pittsburgh Chromatography Forum, Subgroup of the Society For Analytical Chemists of Pittsburgh. SFC-87 had over 200 registered attendees, 14 formal invited presentations, and 17 contributed poster presentations. The meeting was the first ever held that was devoted exclusively to the topic of supercritical fluid chromatography. SFC-87 was truely international in scope with papers presented by scientists from Europe and North America and attendees from Canada, England, France, Germany, Italy, South Africa, Switzerland, and the United States. By any standard of measure, the meeting was a success. I thank the Society For Analytical Chemists of Pittsburgh for providing us with the support to organize and hold SFC-87.

Pittsburgh, PA
September 1988

Curt M. White

Contents

1 Introduction to Composition Gradients in Supercritical Fluid Chromatography

Ernst Klesper and Franz P. Schmitz
Lehrstuhl für Makromolekulare Chemie, Technical University of Aachen, Worringerweg 1, D-5100 Aachen, Federal Republic of Germany

Ernst Friedrich Günter Klesper, born in Cologne, Federal Republic of Germany, received the degree of a Dr. rer. nat. from the University of Hamburg in 1954. He then joined the Wm.T. Burnett Company in Baltimore, Md., as a chief chemist. Thereafter he was a Research Associate at the Johns Hopkins University and Assistant Professor at the University of Maryland. Subsequently, he became associated with the Institut für Makromolekulare Chemie at the University of Freiburg, FRG, as a Wissenschaftlicher Assistent, Dr. rer. nat. habil., Privatdozent, and apl. Professor. *Professor Klesper* now holds a chair for macromolecular chemistry at the Technical University of Aachen. He has a honorary doctorate from the Universidade Federal do Rio de Janeiro and is the author of more than 130 publications. His research interests include the study of reactions on polymers, spectroscopy of polymers, and supercritical fluid chromatography.

Franz Peter Schmitz studied chemistry at the Technical University of Aachen and acquired a doctorate in 1978. Up to the present he is a Wissenschaftlicher Assistent and Privatdozent at the chair for macromolecular chemistry at this university. *Dr. Schmitz* is the author of more than 50 publications and his research interest encompasses the analysis of polymers, including supercritical fluid chromatography.

Summary:

The dependence of the chromatographic parameters, retention, selectivity, plate number, and resolution on the physical parameters, temperature, pressure, and mobile phase composition, is of particular interest in the application of gradient methods in supercritical fluid chromatography (SFC). While this is true also for gas chromatography (GC) and liquid chromatography (LC and HPLC), it is even more relevant for SFC, because the behavior of the chromatographic parameters is more varied in SFC than in either GC or HPLC. It has been known for a long time, for instance, that the capacity ratios may exhibit maxima above the critical temperature T_C depending upon the temperature at constant pressure. A more recent finding is that this also holds true for selectivity, plate number, and resolution.

The variation of the chromatographic parameters on changing the temperature, pressure, and mobile phase composition, is presented with the aid of experimental data obtained in a series of isocratic-isothermal-isobaric separations. A standard chromatographic system, consisting of a mixture of polycyclic aromatic hydrocarbons or styrene oligomers as the analyte, separated on bare silica as the stationary phase in a packed column with different mobile phases, *e. g.* *n*-pentane and CO_2, has been used throughout to allow direct comparison between data. The data are shown in two- or three-dimensional plots and their relevance for gradients is discussed.

Chromatograms obtained by different composition gradients, as well as by a multiple gradients containing a composition gradient, are presented.

Concentrate:

Although the potential of composition gradients is only beginning to be developed, the results obtained so far allow the conclusion that composition programming will supplement pressure or density programming as an alternative method to increase the solvation strength of the mobile phase and to reduce the capacity ratios.

1.1 Introduction

Supercritical fluid chromatography (SFC) is distinguished from gas chromatography (GC) and liquid chromatography (HPLC) by a mobile phase which is both gaseous and solvating. This contrasts with gas chromatography, which uses low pressure gases possessing small or no solvating ability as inert carrier. It also contrasts with liquid chromatography whose mobile phase, although solvating, is present in the liquid state. For a gas to possess dissolution ability for analytes, it must be relatively dense, *i. e.* it must be under higher pressure. Just as important is that the gas should be capable of considerable attractive intermolecular interactions, a property which is not only found to differ widely between liquids but also between gases. Thus different dense gases may show widely different solvation ability, even if their free volume is the same. It is for these two reasons that the use of pressure/density or composition gradients is effective in SFC. Gradients are then created by time-dependent programming of pressure or composition.

With increasing pressure, the density of the supercritical fluid increases and with the latter also the volatility, *i. e.* the speed of transport of the eluite. However, a pressure program has a practical limit when a liquid-like density is reached because the compressibility and the increase in solvation power then become small. The solvation actually may decrease if pressure and density become high enough. Interest has therefore focused on increasing the dissolution power of the supercritical mobile phase not by increasing the pressure, but by the addition of a second component to the supercritical mobile phase. Of particular interest for analytes of widely differing solubility properties is continuous programming of the composition of the mobile phase during a separation, *i. e.* a composition gradient. Addition of a second component, which exhibits a higher solvation for the eluites, is not necessarily associated with as strong a decrease in the free volume of the mobile phase as an equivalent pressure program. At constant temperature and pressure, the second component may therefore be suspected to less adversely affect the interdiffusion coefficient and the viscosity than a pressure gradient at the same temperature, still producing a comparable volatility increase for the eluites.

For a composition gradient, the temperature is also of importance, not only because it determines the vapor pressure of the analyte, as in GC, but also because it changes the intermolecular interactions between the eluite and the gas, as well as between the eluite and the stationary phase, similar to the change in molecular interactions between the eluite and the mobile and stationary phase in HPLC. A higher temperature will increase the interdiffusion coefficient and thereby offsets to some extent the decrease of the interdiffusion coefficient which is expected with the introduction of a second component of higher solvation ability at unchanging pressure. Thus a composition program may be combined to advantage with a positive temperature program. Reviews have already been published about single and multiple gradients in SFC, excepting composition gradients [1], and also about composition gradients [2].

1.2 Composition Gradients

In recent years studies have been extended from mobile phases which consist of only one component, to multiple component mobile phases, particularly binary phases. While the addition of a relatively small amount of a second component to the major component of a mobile phase has been practiced for many years [3], larger numbers of systematic studies on binary mobile phases have only appeared recently. The term modifier has been used for the second component. Apparently two types of modifier action exist: the first consists of interaction of the modifier with the stationary phase, changing its retention properties. For instance, methanol interacts strongly with the active sites of bare silica, particularly those sites which are the most active, strongly reducing the capacity ratio and also the tendency to exhibit tailing. Only small amounts of methanol ($< 1\%$) are needed to obtain this effect. Larger amounts of methanol ($> 5\%$), however, also show the second type of modifier action, because they start to change the solvation properties of the mobile phase considerably. This also influences the retention properties. Therefore, methanol is capable of both types of modifier action. Another kind of modifier action which has also been found involves covalent bonding to the stationary phase. This has been demonstrated, for instance, with 1,4-dioxane as the second component at elevated temperatures [4].

Three-dimensional isobaric network plots are useful in providing an overview with respect to the behavior of the chromatographic parameters capacity ratio, k', selectivity, α, effective plate number, N, and average resolution, R_m*, in dependence on mobile phase composition and temperature [5]. A series of isocratic-isobaric-isothermal chromatograms served to evaluate this behavior on a standard column (25 cm x 0.46 cm) packed with unbonded silica (Lichrosorb 60 or 100, 10 μm), which had previously been exhaustively treated with dioxane. A mixture of naphthalene, anthracene, pyrene, and chrysene was employed as analyte (PAH-1). The average resolution R_m* between the four polycyclic aromatic hydrocarbons has been obtained from the corresponding three individual resolutions R_{ij}* [1]

$$R_{ij}{}^* = \frac{f_{ij}}{g_{ij}} + \frac{d_{ij}}{w_i{}' + w_j{}'} \cdot \sqrt{\ln 4} \qquad (1)$$

where f_{ij} is the depth of the valley between two neighboring peaks as measured from the average peak height, g_{ij} is the average peak height of neighboring peaks as measured from the baseline, d_{ij} is the distance between the baseline intercepts of the two tangents to the neighboring peaks, whereby the tangents are on the sides of the two peaks which oppose each other, and w' the widths of the peaks at half height. **Figure 1** shows such a plot of k' for chrysene, $k'(C)$, for the binary mobile

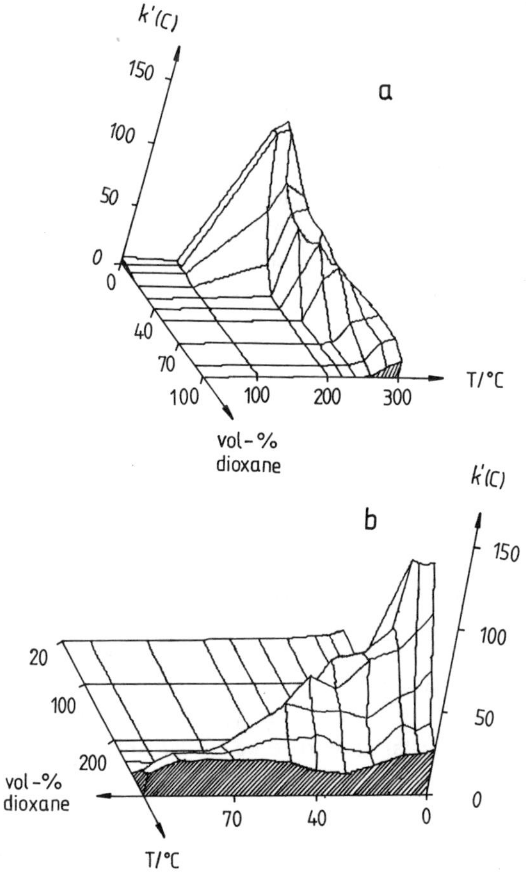

Figure 1
Dependence of the capacity ratio of chrysene, k'(C), on temperature T and mobile phase composition, vol-% dioxane, in two perspectives a and b [5]. The binary mobile phase consisted of a mixture of n-pentane/1,4-dioxane. Stationary phase: Lichrosorb Si 100, 10 μm in a 25 cm x 0.46 cm column. Analyte: PAH-1. Detection: UV at 254 nm. Flow rate at pump (liquid at R. T.): 1 ml/min. Constant outlet resistor (valve of fixed setting and temperature), column outlet pressure, p_e = 20 bar. Reprinted from [5].

phase n-pentane/1,4-dioxane. A strong rise is seen in the region of supercritical temperature, *i. e.* above T_c, which slopes downward strongly to both lower and higher temperatures. The elevation slopes downward also with higher contents of dioxane, *i. e.* with the amount of the more solvating component. Moreover, higher pressure depresses the elevation down in height and volume, as seen in **Figure 2**. This corresponds to the decrease of k' with increasing pressure, as observed for single mobile phases [1]. Thus, the highly solvating second component strongly reduces k'. Elevations above T_c in such three-dimensional plots are also observed for α, N, and R_m*. **Figures 3** and **4** show the selectivity between pyrene and chrysene, α (PC), and the effective plate number of chrysene, N(C). Obviously, the plate number drops strongly with the content of dioxane in the mobile phase, while this is less the case for the selectivity. Therefore, according to eq. (2) the height of the elevation for R_m* depends to varying extents on the contributions of α and N, whenever the amount of dioxane is being changed during a composition gradient

$$R_{ij} = \frac{1}{4} \cdot \frac{\alpha - 1}{\alpha} \cdot \sqrt{N} \qquad (2)$$

$$R_{ij} = R_{ij}^* \text{ if } R_{ij}^* > 1$$

Overall, the resolution tends to decrease with increasing dioxane content. The change in the contributions of α and N to R_{ij} and the clefted appearance of **Figure 4** for N leads in some cases, however, to intermittent maxima in the elevation. This is seen in **Figure 5** for a two-dimensional plot of R_m* *versus* vol-% dioxane [6]. The reason for the intermittent maxima may be connected to an interaction between the dioxane and the silica surface; it is conceivable that molecular layers of dioxane on the surface are responsible.

The behavior of other chromatographic systems with binary mobile phases which have been investigated so far appears to be similar to the system with n-pentane/

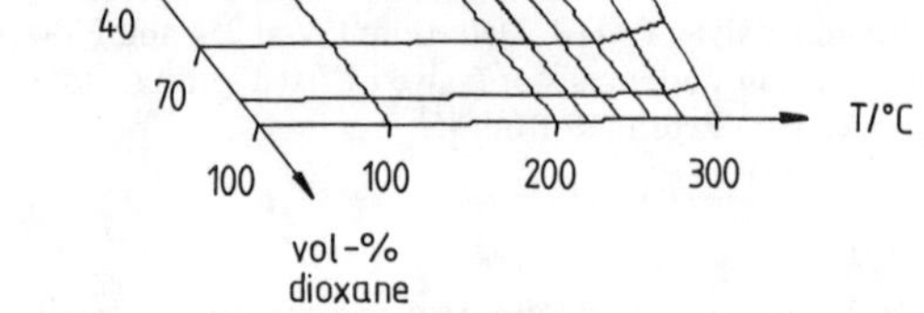

Figure 2
Dependence of the capacity ratio of chrysene, k'(C), on temperature and mobile phase composition, p_e = 36 bar [5]. Other conditions as in Fig. 1. Reprinted from [5].

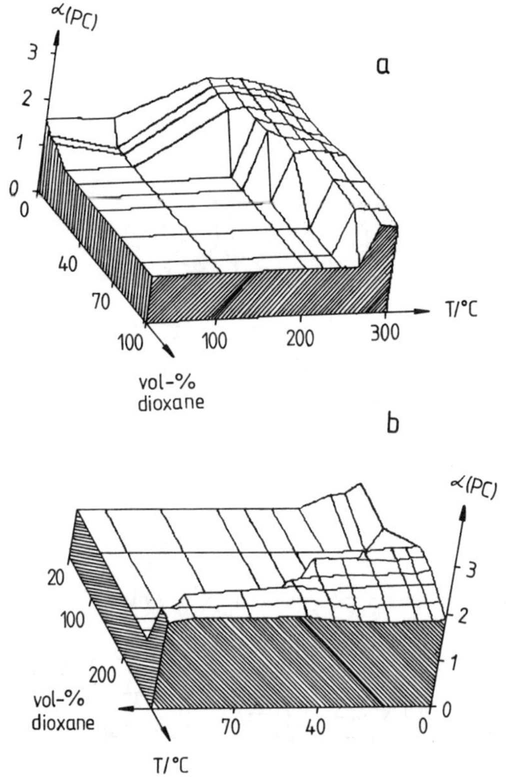

Figure 3
Dependence of the selectivity α between pyrene and chrysene, α(PC), on temperature T and mobile phase composition, vol-% dioxane, in two perspectives a and b [5]. Column outlet pressure, p_e = 20 bar. Other conditions as in Fig. 1. Reprinted from [5].

1,4-dioxane as the mobile phase, silica as the stationary phase, and polycyclic aromatic hydrocarbons (PAH) as the analytes. Thus *n*-pentane/diethyleneglycol dimethyl ether [7], ethane/1,4-dioxane [8], and CO_2/1,4-dioxane [8] have been employed with bare silica and PAH. Similar results have also been obtained with CO_2/alcohol, whereby methanol, ethanol, and 2-propanol were used in conjunction with bare silica or bonded phases and with different analytes [9]. With a constant modifier content in the mobile phase the same study led to the separation of enantiomeric phosphine oxide compounds on a Pirkle phase.

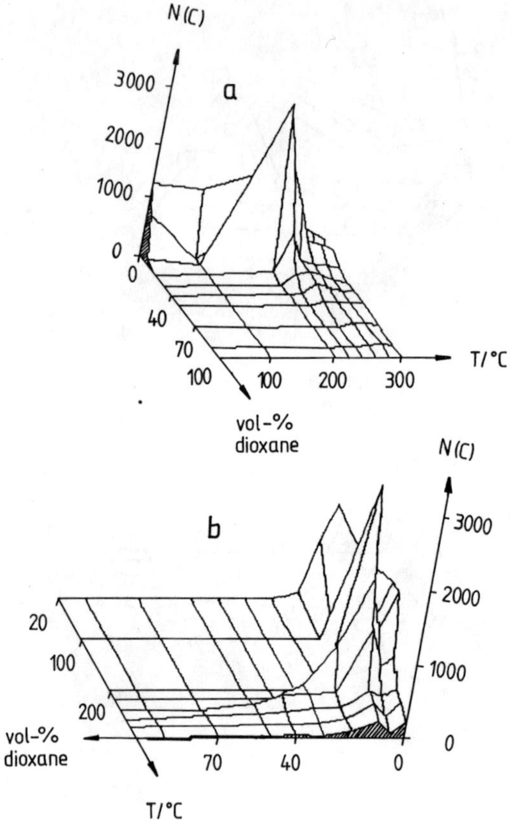

Figure 4
Dependence of the effective plate number of chrysene, $N(\mathrm{C})$, on temperature T and mobile phase composition, vol-% dioxane, in two perspectives a and b [5]. $p_e = 20$ bar. Other conditions as in Fig. 1. Reprinted from [5].

In contrast to mobile phases containing an unchanging modifier content, the programming of the modifier content was demonstrated relatively late [10–13]. This is also in contrast to pressure [14, 15] and pressure-temperature [16, 17] programming, for which examples appeared earlier in the literature. Suggestions as to the potential utility of gradient elution in SFC can be found, however, also rather early in the literature [18, 19].

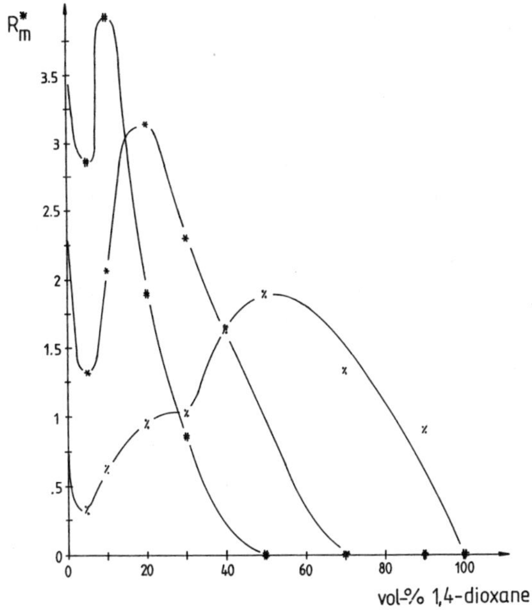

Figure 5
Variation of R_m^* with the content of dioxane in the binary mobile phase of *n*-pentane/1,4-dioxane at a constant column outlet pressure p_e = 36 bar [6]. Temperatures: (#) 240 °C, (*) 260 °C, (%) 300 °C. Other conditions as in Fig. 1. Reprinted from [6].

Pressure or composition gradients are particularly useful for analytes whose individual eluites differ widely in solubility and/or adsorption. Because oligomers are usually composed of individual species of widely differing molecular weight, oligomer separation has become a domain of SFC since the work of *Jentoft* and *Gouw* [14] who separated styrene oligomers by means of a pressure gradient. Pressure gradients or pressure-temperature gradients have been the only gradient procedures applied until it was demonstrated that gradients of eluent composition were also applicable just like in HPLC. The similar effect of pressure and composition gradients is not surprising, because both chromatographic methods are based on the same transport principle. Owing to the dense nature of the mobile gas phase, the eluite molecules are transported along the separation column wholly or partially by the molecular interactions with the mobile phase, *i. e.* by solubility, and not, or not to the same extent, by vapor pressure as in GC.

For the separation of oligomers of styrene and related monomers, the following binary mobile phases have been found suitable: pentane/cyclohexane [11], diethyl ether/1,4-dioxane [11], pentane/dioxane [11, 20, 21], and propane, butane, or hexane with dioxane [11]. In these eluent systems, cyclohexane or dioxane is the component with the higher solvent strength for the vinyl arene oligomers. Increasing concentration of either cyclohexane or dioxane therefore enabled the elution of higher molecular-weight species. **Figure 6** shows oligostyrene separations using alkane/dioxane mixtures [11]. The composition gradient is shown with the chromatograms. In Figure 6a propane/dioxane and in Figure 6b hexane/dioxane has been employed as the binary mobile phase. It is straightforward to change the composition program to expand or compress portions of the chromatograms as desired. In Figure 6a, for in-

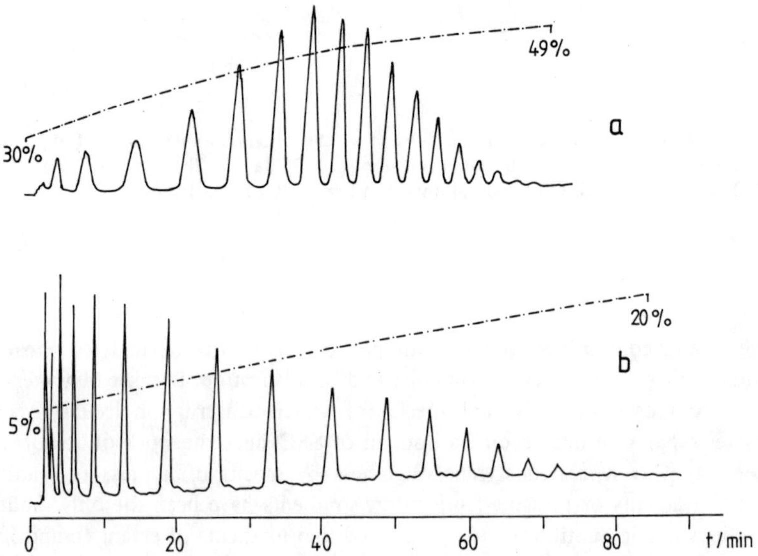

Figure 6
Chromatograms of a styrene oligomer, PS 800, as the analyte with (a) propane/dioxane and (b) hexane-dioxane as the binary mobile phase [11]. Content (vol/vol) of dioxane in the eluent mixture during the program [– · – · – ·]. Column temperatures: (a) 240 °C, (b) 275 °C. Column exit pressures p_e at the start of the chromatograms: (a) 80 bar, (b) 41 bar. Other conditions as in Fig. 1. Reprinted from [2].

stance, the first portion of the chromatogram is expanded, while in Figure 6b it is primarily the middle part.

Of the eluent systems, particularly the combination pentane/dioxane has been frequently used. With styrene oligomers, separations of individual oligomeric species were obtained up to a molecular weight range of 5500 g/mol, and even polystyrene samples could be separated [10] with molecular weights up to 600000 g/mol as shown in **Figure 7**. However, in the latter case individual oligomeric species could not be separated. The separation of samples of polymers is, of course, limited by the requirement that the molecular weight distributions of samples neighboring in molecular weight are not to overlap. As expected, higher dioxane contents and higher pressures were generally needed for eluting the polymer samples as compared to the oligomeric species. In Figure 7a, several polystyrene standards of lower

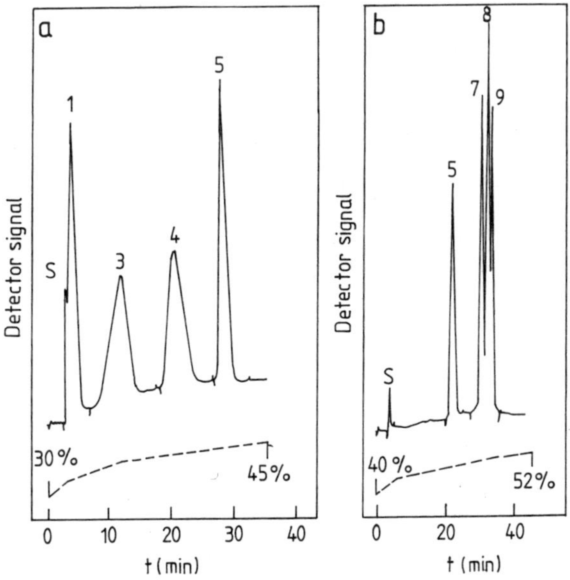

Figure 7
Separation of polystyrenes in n-pentane/1,4-dioxane at 250 °C and initial p_e = 166 bar [10]. Program of dioxane content (vol/vol) [– – –]. S, sample solvent; 1, polystyrene sample (PS) with number-average molecular weight $\overline{M}_n$ = 3,600 g/mol; 3, PS with weight-average molecular weight $\overline{M}_w$ = 10,000 g/mol; 4, PS with $\overline{M}_w$ = 53,700 g/mol; 5, PS with $\overline{M}_w$ = 93,000 g/mol; (b): 7, PS with $\overline{M}_w$ = 254,000; 8, PS with $\overline{M}_w$ = 392,000; 9, PS with $\overline{M}_w$ = 600,000. Reprinted from [10].

molecular weights are separated; Figure 7b shows standards of higher molecular weight. Pentane/1,4-dioxane and CO_2/1,4-dioxane gradients have also been employed to separate other vinyl arene oligomers [22]. Additionally, the separation of oxyethylene oligomers has been demonstrated by *Gere* [23] using CO_2/methanol as the mobile phase and a column packed with polystyrene. This latter eluent pair has also been used for separations of aromatic hydrocarbons by gradient elution [13].

Figure 8 shows chromatograms of an anionically prepared sample of 4-vinylbiphenyl oligomers at two different temperatures, the mobile phase being pentane/dioxane. The quality of this separation clearly demonstrated that at least two homologous series were obtained with the anionic oligomerization reaction [20],

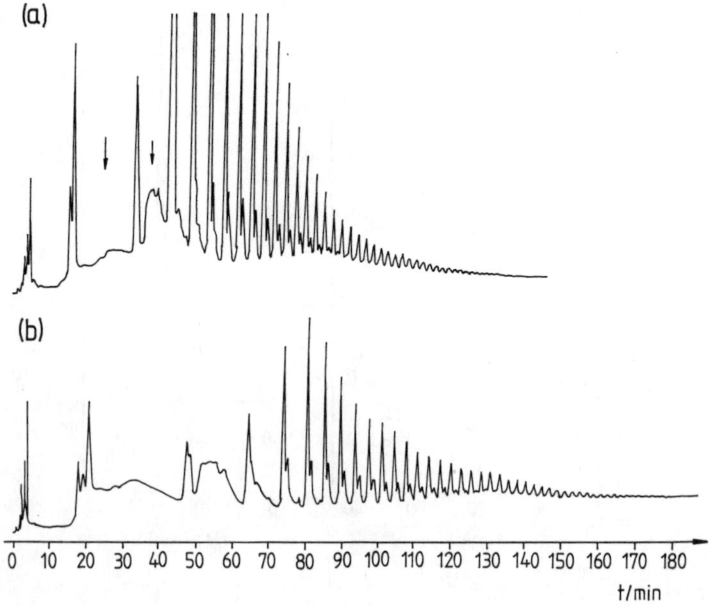

Figure 8
Chromatograms of 4-vinylbiphenyl oligomers [20]. Temperatures, (a) 250 °C; (b) 270 °C. Column inlet pressure at the start of the chromatograms, (a) 49 bar; (b) 45 bar. Gradients: 5 to 60% dioxane in pentane (vol/vol) within 220 min according to eq. (3). Other conditions as in Fig. 1. Reprinted from [20].

and a comparison could be made between radical and anionic oligomerizations by means of SFC [21]. The eluent composition gradient applied for these separations was formed by linearly increasing the ratio Q_B with time,

$$Q_B = \frac{P_B}{1 - P_B} \tag{3}$$

where P_B is the volume fraction of the second, more solvating component B, as measured at the pumps in the liquid state at ambient temperature. During a composition program, the column pressure increases, if the second component has a higher viscosity than the first component and if the resistor at the column exit, *i. e.* the flow resistance, is kept at a fixed setting and temperature and is thereby kept invariable. Such a pressure increase is not necessarily dramatic, but still enhances the effect of the eluent composition gradient; the main effect on k', however, is due to the increasing content of the second component in the mobile phase [11].

1.3 Multiple Gradients Containing Composition Gradients

The characteristic dependence of the chromatographic parameters k', α, n, N, R, and R_m^* on the physical parameters temperature, pressure, density, and eluent composition leads to the question of how the separation with respect to analysis time or to resolution may be improved by multiple gradients. With simultaneous multiple gradients, two or more of the physical parameters are programmed simultaneously. Some separations quoted in the previous section have, in fact, been the result of a multiple program, *i. e.* of a combined composition-pressure gradient, but there the pressure program was only of a modest relative pressure increase and was due to the composition program by virtue of a rising viscosity and an invariant resistor at the column exit.

N-Vinylcarbazole oligomers, which were found difficult to elute by composition programming alone, were successfully separated on a packed silica column in pentane/dioxane by combining gradients of pressure and eluent composition [20]. Several alkylphenones were separated on a capillary column in CO_2/2-propanol by such a composition-pressure-program [24]. A knowledge of the dependences of k' and R on T, p, and % composition will apparently be particularly helpful in designing an optimum multiple program. For obtaining a maximum in resolution during gradient elution, it will be of advantage to remain at the crest of the elevation for R during composition or composition-pressure programming. Because the crest is moving to higher temperatures with higher contents of dioxane in the pentane-dioxane mobile phase and also with higher pressures, a positive composition program may be combined with a suitable positive temperature program, *i. e.* a composition-temperature-program or a composition-pressure-temperature-program is then employed. This has the prerequisite that the chromatographic system shows a behavior basically similar to that in Figures 3 and 4, *i. e.* an elevation for α, N, and R above T_c.

For choosing a suitable temperature program which is to be combined with a composition program, it is helpful to plot the temperatures of the maxima of resolution *versus* the composition, with pressure as the parameter. A corresponding plot is seen in **Figure 9** for the binary mobile phase CO_2/1,4-dioxane, a packed silica gel column, and PAH-1 as the analyte [25].

In **Figure 10**, two separations of oligostyrene PS-800 by the same simultaneous composition-pressure gradient are seen. The chromatogram in Figure 10a was obtained under isothermal, and that in Figure 10b under temperature-programmed

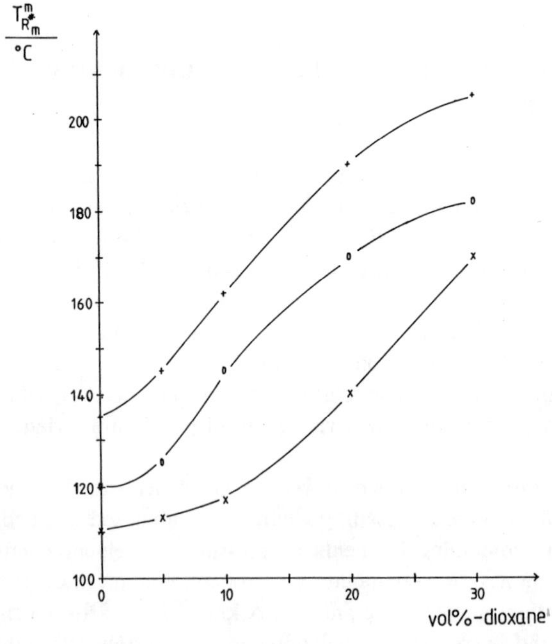

Figure 9
Dependence of the temperature for the maximum of mean resolution, $T^{m}_{R_m}$, on the composition of the binary mobile phase CO_2/1,4-dioxane and the column exit pressure, p_e [8, 25]. Key: (x) p_e = 150 bar; (o) p_e = 200 bar; (+) p_e = 250 bar. Other conditions as in Fig. 1. Reprinted from [2].

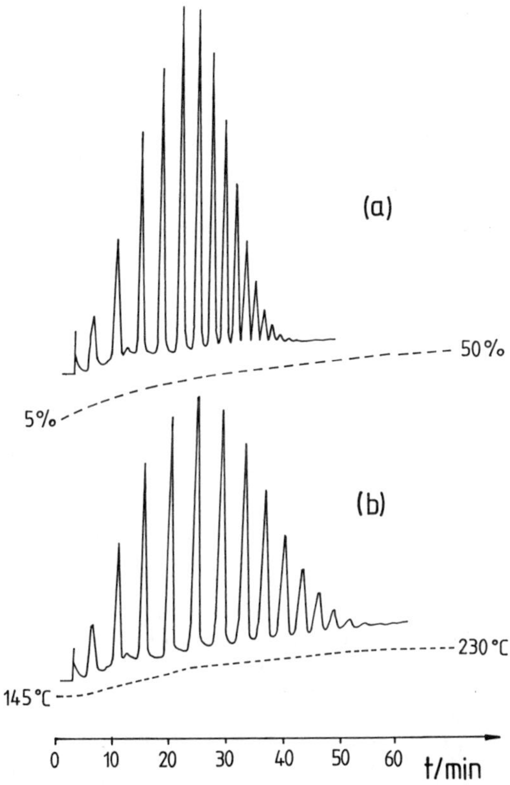

Figure 10
Chromatograms of PS-800 as the analyte and CO_2/1,4-dioxane as the binary mobile phase. (a) Constant temperature at T = 145 °C, composition program from 5 to 50 vol-% dioxane, pressure program from 250 to 295 bar. (b) Analogous to (a) but with temperature program from 145 to 230 °C. Other conditions as in Fig. 1. Reprinted from [2].

conditions. The temperature program for Figure 10b was determined using the graph shown in Figure 9, *i. e.* the temperature for maximum resolution was taken from the $T^{m}_{R_m}$-plots. The resulting temperature program was applied simultaneously with the composition-pressure program. Comparison between the two chromatograms shows that the distance and the resolution between the oligomer peaks is increased on superimposing a temperature program. It should be noted, however, that Figures 9 and 10, although valid for the same chromatographic system, have been obtained for the analytes PAH-1 and for lower fractions of PS 800, but not for whole PS-800.

Acknowledgment

We thank the Arbeitsgemeinschaft Industrieller Forschungsvereinigungen for financial support. Thanks are also due to the publishers of Journal of Chromatography (Figs. 1–6, 8–10) and Polymer Communications (Fig. 7) for permission to reprint figures.

References

[1] *E. Klesper* and *F. P. Schmitz*, J. Chromatogr. **402** (1987) 1–39.
[2] *F. P. Schmitz* and *E. Klesper*, J. Chromatogr. **388** (1987) 3–21.
[3] *T. H. Gouw* and *R. E. Jentoft*, Adv. Chromatogr. **13** (1975) 1–40.
[4] *F. P. Schmitz, D. Leyendecker*, and *D. Leyendecker*, J. Chromatogr. **389** (1987) 245–250.
[5] *D. Leyendecker, F. P. Schmitz, D. Leyendecker*, and *E. Klesper*, J. Chromatogr. **393** (1987) 155–167.
[6] *D. Leyendecker, D. Leyendecker, F. P. Schmitz*, and *E. Klesper*, J. Chromatogr. **392** (1987) 101–122.
[7] *D. Leyendecker, F. P. Schmitz*, and *E. Klesper*, Chromatographia **23** (1987) 171–178.
[8] *D. Leyendecker, D. Leyendecker, F. P. Schmitz, B. Lorenschat*, and *E. Klesper*, J. Chromatogr. **398** (1987) 105–123.
[9] *D. Mourier, P. Sassiat, M. Caude*, and *R. Rosset*, J. Chromatogr. **353** (1986) 61–75.
[10] *F. P. Schmitz* and *E. Klesper*, Polymer Commun. **24** (1983) 142–144.
[11] *F. P. Schmitz, H. Hilgers*, and *E. Klesper*, J. Chromatogr. **267** (1983) 267–275.
[12] *F. P. Schmitz* and *E. Klesper*, Makromol. Chem. Rapid Commun. **2** (1981) 735–739; Polym. Bull. **5** (1981) 603–608.
[13] *A. L. Blilie* and *T. Greibrokk*, J. Chromatogr. **349** (1985) 317–322.
[14] *R. E. Jentoft* and *T. H. Gouw*, J. Pol. Sei., Polym. Lett. Ed. **7** (1969) 811–813.
[15] *G. W. A. Rijnders*, in "Column Chromatography" 5th International Symposium on Separation Methods, Sauerländer, Aarau (1970) pp 192–200.
[16] *W. Hartmann* and *E. Klesper*, "Airapt International Pressure Conference", Boulder, Colorado (1977) preprint K-6-C.
[17] *E. Klesper* and *W. Hartmann*, Eur. Polym. J. **14** (1978) 77–88.
[18] *N. M. Karayannis* and *A. H. Corwin*, Anal. Biochem. **26** (1968) 34–50.
[19] *J. E. Conaway, J. E. Graham*, and *L. B. Rogers*, J. Chromatogr. Sci. **16** (1978) 102–110.
[20] *F. P. Schmitz, H. Hilgers, B. Lorenschat*, and *E. Klesper*, J. Chromatogr. **346** (1985) 69–79.
[21] *F. P. Schmitz* and *H. Hilgers*, Makromol. Chem., Rapid Commun. **7** (1986) 59–63.
[22] *F. P. Schmitz, H. Hilgers*, and *B. Gemmel*, J. Chromatogr. **371** (1986) 135–144.
[23] *D. R. Gere*, Appl. Note 800-3, Publ. No. 43-5953-1690, Hewlett-Packard Co., Avondale, PA, 1983.
[24] *C. R. Yonker* and *R. D. Smith*, Anal. Chem. **59** (1987) 727–731.
[25] *D. Leyendecker, D. Leyendecker, F. P. Schmitz*, and *E. Klesper*, HRC & CC **9** (1986) 525–527.

2 Recent Advances in Packed Column SFC

Frank Pacholec, David S. Boyer, Raymond K. Houck, and Athos C. Rosselli
Suprex Corporation, SFC Research Center, 125 William Pitt Way, Pittsburgh, PA 15238, USA

Frank Pacholec is a native of Hamtramck, Michigan. He received a B. S. in Chemistry and Biology in 1978 from the University of Michigan, and a Ph. D. from Wayne State University in 1983 under the direction of *Professor Colin F. Poole*. In 1983, he joined Monsanto Co. as a research chemist, prior to becoming Manager of Technology of Suprex Corp. in 1986. He is presently employed in the Detergents and Phosphates Division of Monsanto Co. at their World Headquarters in St. Louis, MO. *Frank* has over twenty publications in areas including gas chromatography, liquid and ion chromatography, high performance thin-layer chromatography, and supercritical fluid chromatography. He and his co-workers are credited with the first report documenting the coupling of modern ion chromatography and mass spectrometry.

Athos C. Rosselli is a Research Chemist for Suprex Corporation of Pittsburgh, PA, manufacturers of supercritical fluid chromatographs. He received his Bachelor's degree in chemistry from New York University in 1974. Later that year he came to Pittsburgh to attend graduate school in organic chemistry at Duquesne University until 1977. In that year he became a quality-control chemist for Pennex Products in Verona, PA, where he performed GC and LC analyses of over-the-counter preparations. In 1978 he accepted a position with the Chemical Plants Division of Dravo Corporation in Pittsburgh as an Assistant Engineer. His responsibilities there included physical and chemical testing and the development of analytical GC techniques for residual solvent determination in a proprietary solvent extraction process. He was also responsible for the generation of design parameters for a diatomite ore extraction pilot plant. Early in 1982 he became Acting Laboratory Supervisor for the Process Engineering Department where he directed a crash laboratory program to provide information for the design and construction of a solvent extraction pilot plant. In late 1983 *Mr. Rosselli* became a Senior Chemist with Pittsburgh Testing Laboratory (now PTL-Inspectorate), also based in Pittsburgh. At PTL he developed a variety of analytical methods for the determination of PCB's, pesticides and herbicides in various matrices. He also directed various special project including a program for the validation of inorganic and organic vapor dosimeters. In 1986 he came to Suprex where he is currently in charge of customer applications and technical support.

Raymond K. Houck is Chairman and founder of Suprex and has been active in SFC application and technical development for five years. He has experience in supercritical fluid extraction and GC from work at Carnegie-Mellon University and Air Products & Chemicals, Inc. He has a MSIA from Carnegie-Mellon University and a BSChE from Pennsylvania State University.

Summary:

Nearly all of the classical and pioneering work in supercritical fluid chromatography (SFC) was performed on packed columns, including the first experimental description of the use of a supercritical fluid as a chromatographic mobile phase in 1962 by *Klesper*, *Corwin*, and *Turner*. More recently, however, the use of open-tubular (capillary) columns has dominated the SFC literature. This paper reviews some of the recent advances made in packed column SFC, and explores the present limitations and future potential of packed columns in the practice of SFC.

A detailed description of the physical characteristics and dimensions of modern packed SFC columns is presented, including column length, column diameter, and particle size. The restrictions placed on these dimensions by various SFC hardware

components such as the pumping system and choice of detector is discussed. The potential merit of modified CO_2 and other mobile phases is reviewed, also with regard to the limitations of current SFC hardware and detector technology.

One of the biggest issues surrounding the use of packed columns in SFC is their high surface activity. The origin of this activity is discussed, and a variety of approaches that attempt to alleviate this situation are presented. The elution densities of homologous series of both *n*-alkanes and silanol-terminated polysiloxanes are measured and compared on packed and open-tubular columns.

The sensitivity and reproducibility typically obtainable with packed column SFC is reviewed. A "sensitivity enhancement" system is introduced that extends the maximum volume that may be introduced to a packed column of a given diameter. Using this system, it should be feasible in the future to lower detection limits attainable with packed column SFC/FID to the 10–50 ppb range.

The first direct comparison of resolution with packed and open-tubular columns is presented for a series of *n*-alkanes. Trennzahl numbers are calculated and compared, followed by a detailed discussion that presents potential explanations of the surprising results. Finally, the current and future areas of SFC development are discussed.

Concentrate:

Some of the important recent developments in packed column SFC technology are reviewed. Topics included are physical column dimensions, mobile phases and modifiers, column inertness, sensitivity, reproducibility, and resolution. Packed columns clearly have unique features that establish their importance in the modern practice of SFC. Data are presented to illustrate both the utility and limitations of current packed column technology, with a look forward to future developments.

2.1 Introduction

Packed columns have been used extensively to achieve a wide variety of separations in gas chromatography (GC), liquid chromatography (LC), and supercritical fluid chromatography (SFC). The first published report of SFC by *Klesper, Corwin,* and *Turner* in 1962 employed a 30 in. long column packed with 33% Carbowax 20M on Chromosorb W to separate a mixture of metal porphyrins [1]. In fact, during the time period from 1962 to 1981 nearly all of the publications related to SFC (totalling $\approx$ 90) involved the use of packed columns [2]. In 1981, *Novotny, Springston, Peaden, Fjeldsted,* and *Lee* published an article in Analytical Chemistry describing the use of open-tubular wall-coated capillary columns with SFC [3]. Since then, the

SFC literature has been dominated by reports dealing with the use of open-tubular columns, while packed column SFC has experienced limited use [2].

The preferential use of capillary columns in SFC may be due in part to the currently held belief in GC that the present state-of-the-art and future of the technique lies with the use of high resolution capillary (open-tubular) columns [4]. In 1979, *Walter Jennings* published comparative separations of peppermint oil which he called "the three generations of gas chromatography" [5]. This series of separations, which has been reproduced in **Figure 1**, was performed on a 6 ft. x 0.25 inch i. d. packed column, a 500 ft. x 0.03 inch i. d. stainless steel open-tubular column, and

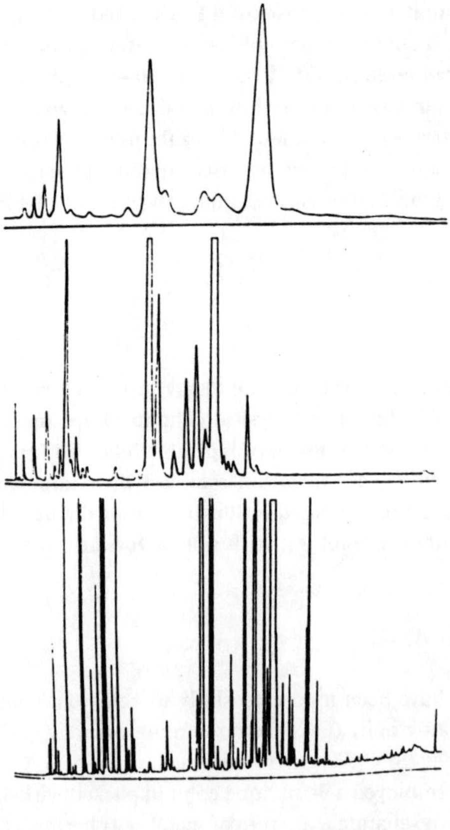

Figure 1
Three generations of gas chromatography. Peppermint oil separated on (upper) 6 ft x 0.25 in i. d. packed column, (middle) 500 ft x 0.03 in i. d. stainless steel open-tubular column, and (lower) 50 m x 0.25 mm i. d. glass open-tubular column. All columns contained Carbowax 20M stationary phase and were operated under optimized conditions. (Reprinted with permission from reference [5].)

on a 50 m x 0.25 mm i. d. glass open-tubular column, each containing the same stationary phase, Carbowax 20M. This series of separations serves to illustrate the dramatic increase in resolution obtained in going from traditional packed columns, to earlier generation stainless steel open-tubular columns, and finally to modern glass open-tubular columns. The packed column chromatogram contains fewer than fifteen peaks, while the separation performed on the glass open-tubular column contains 8 – 10 times as many peaks. This comparison and others like it have served to instill into gas chromatographers the superior nature of capillary columns, and this belief has been carried over to SFC. Liquid chromatographers on the other hand may tend to have fewer preconceived notions regarding packed *vs.* open-tubular SFC due to the fact that the poor diffusion rates of solutes in liquids has largely limited the use of open-tubular columns in LC.

We sought to make a similar separation comparison of packed and open-tubular columns in SFC. For the experiment, columns typical of those currently used in SFC were selected: a 10 cm x 1 mm i. d. octadecylsilane packed column (5 μm particle size), and a 10 m xm 50μm i. d. open-tubular DB-5 column, 0.2 μm film thickness (J & W Scientific, Folsom, CA, USA). A series of *n*-paraffins ranging in carbon number from 12 to 44 were used as solutes, and each separation was optimized to produce the highest resolution attainable in approximately thirty minutes. The resulting chromatograms are shown in **Figure 2**. When compared to the dramatic differences in resolution between packed and capillary column chromatograms in GC, these SFC chromatograms are not only more similar, but almost identical. This separation comparison will be discussed in greater detail in the sections to follow, and is initially presented here to introduce the notion that packed columns may deserve a great deal more attention than they are currently receiving.

2.2 Physical Characteristics and Dimensions of Modern Packed SFC Columns

2.2.1 General

In the course of development of packed column SFC, a wide variety of columns have been implemented. However, the types of packed columns employed in the modern practice of SFC are almost identical to those used in modern high performance liquid chromatography (HPLC). The types of sorbents used in both techniques are the same, while the lengths and internal diameters of SFC packed columns are more limited than in HPLC due to pressure drop considerations and the upper flow limit of some commonly used SFC detectors.

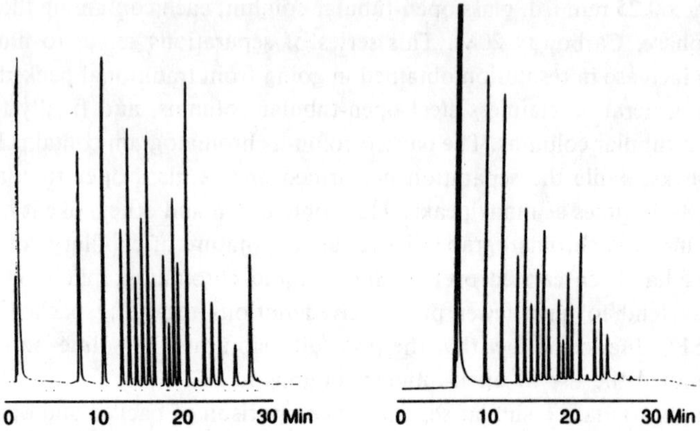

Figure 2
Packed (left) *vs.* open-tubular (right) separation of standard mixture of C_{12} to C_{44} normal paraffins. Conditions optimized in each case for highest resolution in less than ≈ 30 minutes.

As in HPLC, the choice of sorbent must be compatible with the supercritical fluid mobile phase employed. For carbon dioxide, the most commonly used SFC mobile phase, this includes both "reversed phase" and "normal phase" silica based packings (see **Table 1**). Polymeric packing materials based on polystyrene divinylbenzene copolymers may also be employed. One of the serious problems with all of these packing materials is their high activity, due in the case of the silica-based columns to the inevitable existence of some unbonded "bare" Si-OH sites. While packed columns based on polystyrene divinylbenzene copolymers do not contain unbonded silica sites, they can contain metal impurities, and/or polar compounds remaining from the polymerization process or from oxidation of the sorbent surface [6]. The resulting high activity can lead to peak tailing, irreversible adsorption, or high elution densities for polar analytes. This issue will be discussed in detail in a later section dealing specifically with column inertness.

2.2.2 Particle Diameter and Column Length

Packed columns currently in use in SFC contain packing materials having particle diameters 5 μm – 20 μm or larger, and are typically 5 cm – 20 cm in length. The particle diameter and length need to be considered together as they are involved in determining the pressure drop across the column as seen in eq. (1) [7].

Table 1
List of various types of packing material used in packed column SFC.

SFC Column Packing Materials
Reversed Phase
octadecylsilane
octyl
methyl
ethyl
Normal Phase
alumina
amino
cyano
diol
silica
Other
polystyrene-divinylbenzene based
GC stationary phases, bonded and cross-linked

$$P = \frac{\bar{u}\,(\psi)\,(n)\,(L)}{d_p^2} \tag{1}$$

where u = average linear velocity, (cm/s), ψ = constant (≈ 1000 for spherical porous packing), n = viscosity (poise), L = column length (cm), d_p = particle diameter of packing (μm).

Currently there is controversy regarding the consequences of a high or low column pressure drop in terms of resolution and selectivity. At the time of writing, it appears that a column pressure drop equal to 5–10% of the total column head pressure has little or no ill effect on selectivity, and according to one theory may actually enhance resolution, as will be discussed later. At higher column pressure drops, the densities at the front and back ends of the column are very different, and because mobile phase solvating power is directly proportional to density, considerable eluting power can be lost across the column. In practice, a good practical limit in length for a column packed with 10 μm particles is about 25 cm. One interesting combination that has not as yet been thoroughly investigated is the use of longer length packed columns (*i. e.* 15 cm to 1 meter), containing pellicular packing materials of particle diameter 30 μm or larger. Some very preliminary results obtained with these types of columns have looked promising [8]. If they are capable of producing reasonably good resolution and peak width, columns of this type might find widespread use due to their ease of preparation.

2.2.3 Column Internal Diameter

For packed columns, pressure drop and resolution are essentially independent of column internal diameter (if a similar column packing efficiency is assumed for different diameter columns). Therefore, the selection of column diameter is dictated by the cost of the mobile phase, the volume flow capability of the SFC pumping system, the type of detector being used, and the required level of sensitivity of the analysis.

The cost of the mobile phase is generally of no concern when the more common mobile phases such as carbon dioxide, nitrous oxide, or pentane are used. With these mobile phases, SFC is generally less expensive to operate than reversed-phase HPLC.

The second consideration is the pumping capacity of the SFC hardware. The pump must be able to deliver a pulseless flow of mobile phase at a volume flow rate sufficient to provide a linear velocity that will generate narrow peaks, without the need to refill the pump so often as to be a nuisance.

Figure 3 contains some calculations made under typical operating conditions for a 1.0 mm i.d. packed column. The volume flow rate of carbon dioxide at the outlet of the column under these conditions is approximately 104 ml/min (expanded at ambient pressure). This corresponds to approximately 178 μl/min of liquid carbon dioxide. Under these conditions, approximately 85 ml of liquid CO_2 would be consumed in an eight-hour work day. Therefore a 200 ml syringe pump would last approximately 19 hours, a reasonable amount of time, even if an autosampler is used overnight. On the other hand, a 2.0 mm i.d. packed column operating under similar conditions would empty the syringe pump in approximately 4.7 h, making the filling procedure more of a consideration.

The level of sensitivity required for a specific analysis can also influence the selection of the column internal diameter. It should be noted that the lower limit of the column internal diameter depends on the smallest diameter column that can be packed in a manner as to produce acceptable column efficiency. Narrower diameter columns, however, have less total sample and sample volume capacity than columns of larger diameter. Therefore if maximum sensitivity is required for an analysis, the largest diameter column that can be packed efficiently, and operated with acceptable chromatographic efficiency at a total volume flow rate compatible with the detector in use is the column of choice. It is this last consideration (the detector) that is the single most important factor influencing the selection of column internal diameter.

There are two basic types of SFC detectors: the flow-through type where detection occurs in the supercritical or liquid state in a flow-through cell, and ionization detectors, where the column effluent expands to ambient or near-ambient pressure just prior to entering the detector and subsequent combustion, ionization, *etc.*

PACKED COLUMN CALCULATIONS: 1.0 MM ID

A. PARAMETER ASSUMPTIONS

1. Average column pressure = 360 atmospheres
2. Column pressure drop = 8 atmospheres
3. Inlet pressure = 364 atmospheres, Outlet pressure = 356 atmospheres
4. Column temperature = 45 °C
5. Average column density = 0.925 g/cm^3
6. Column density drop = 0.004 g/cm^3
7. Inlet density = 0.927 g/cm^3, Outlet density = 0.923 g/cm^3
8. Particle diameter = 10 micrometer
9. Column length = 10 cm (100 mm)
10. Hold-up time = 0.25 minutes = 15 seconds

B. CALCULATIONS:

1. Average linear velocity

$$\bar{u} = \frac{[\ 10 \text{ centimeter}\ ]}{[\ 15 \text{ seconds}\ ]} = 0.67 \text{ cm/sec}$$

2. Volume flow rate - Supercritical fluid

$$F = [\ 0.67 \text{ cm/s}\ ][\ 3.14\ ][\ 0.050 \text{ cm}\ ]^2[\ 0.60\ ][60 \text{ s/min}]$$

$$= 0.190 \text{ cm}^3\text{/minute}$$
$$= 190 \text{ microliters/minute}$$

3. Mass flow rate

$$m = [\ 0.925 \text{ grams/cm}^3\ ][\ 0.190 \text{ cm}^3\text{/min}\ ]$$

$$= 0.176 \text{ grams/minute}$$

4. Molar flow rate

$$M = \frac{[\ 0.176 \text{ g/min}\ ]}{[\ 44.0 \text{ g/mole}\ ]} = 0.00399 \text{ moles/min}$$

5. Expanded gas volume flow

$$F_g = [0.00399 \text{ moles/min}\ ][\ 22{,}400 \text{ cm}^3\text{/mole}\ ][318/273]$$

$$= 104 \text{ cm}^3\text{/minute} \quad \text{(into FID or MS)}$$

Figure 3
Linear velocity, volume flow rate, mass flow rate, molar flow rate, and expanded gas volume flow are calculated for a typical set of operating conditions with a 1.0 mm i. d. packed column. The parameter assumptions made for the calculations are also given.

In the first type, the column internal diameter must be large enough to accommodate a volume flow rate at which the detection cell does not contribute significantly to band broadening. The two most commonly used SFC detectors fitting into this category are the ultraviolet (UV) absorption detector, and the fluorescense detector. Excessive band broadening within the detector cell is generally not a problem with large i. d. columns that are compatible with larger cell volumes, but can be an important consideration with smaller i. d. columns. As a point of reference, for a packed column having a column internal diameter of 1.0 mm, operating at a linear velocity of $\approx$ 0.2 cm/s, initial results indicate that a 10–12 μl UV cell of path length 1.0 cm contributes only modestly to total peak band broadening ($\approx$ 5–10%) [8].

With the ionization detectors there is generally never a problem with too little flow, as these detectors accept the mobile phase after it has expanded to ambient or near-ambient conditions. Excessively high flow rates, however, can cause complications. As the supercritical fluid mobile phase expands to ambient pressure, a large Joule-Thompson cooling effect occurs due to adiabatic expansion. For flame-based detectors such as the flame ionization detector (FID), this expansion and cooling can cause either decreased ionization efficiency in the flame, or extinguish it entirely. For this reason, it is common to place the end of the restrictor about one to two inches below the FID flame jet when using packed columns. This serves to keep the flame more stable during pressure or density programming, and also allows higher total flow rates to be introduced into the FID without extinguishing the flame. Even then, the largest practical internal diameter packed column that can be used in conjunction with an FID is $\approx$ 1.0 mm. For ionization detectors such as the electron-capture detector (ECD) that exhibit an extremely temperature-dependent response, this phenomenon can cause baseline drift so severe as to render the system useless. Although packed columns have not as yet been successfully interfaced to an ECD or a nitrogen-phosphorus detector (NPD), they have been shown to work well with mass spectrometry using ion-beam type interfaces [9].

2.3 Mobile Phases and Modifiers

Carbon dioxide (CO_2) is by far the most commonly used SFC mobile phase. There are several reasons for this. The critical temperature and critical pressure of CO_2 are 31.1 °C and 73.8 atm respectively. Carbon dioxide can thus be used to elute thermally labile molecules at mild temperatures, while operating at pressures that are not beyond the capabilities of modern high pressure SFC mobile phase delivery systems. Carbon dioxide also does not respond to the flame ionization detector (FID), and is non-toxic and non-flammable. The most significant shortcoming of CO_2 as a mobile phase is its limited ability to elute polar molecules. Carbon dioxide is non-polar, though it is polarizable. With supercritical CO_2 at typical SFC operating conditions (70–450 atm), the range of δ (Hildebrand solubility parameters) achieved overlaps with the solubility parameters of several common organic solvents such as freons, pentane, methylene chloride, THF, toluene, and methyl and isopropyl alcohol [10]. Though the Hildebrand solubility parameter is at best only a first approximation to the actual physical phenomena involved [3], it provides the SFC user with the useful rule of thumb that if a sample has appreciable solubility in one or more of the above-mentioned organic solvents, there is a reasonable chance that it will migrate in supercritical CO_2 This obviously excludes many polar molecules having zero or limited solubility in organic solvents that are of great interest to the pharmaceutical and biotechnology industries.

There is an additional problem encountered when attempting to elute even moderately polar compounds from packed columns. Even when supercritical CO_2 is capable of mobilizing a particular species, the compound can become irreversibly adsorbed onto the column due to the high level of activity of most LC (and SFC) sorbents. Therefore, if CO_2 is to be successfully used to elute molecules similar in polarity to those eluted by CO_2 with inert fused silica open-tubular columns, the surface activity must be decreased. One approach attempting to accomplish this has involved the use of modifiers.

Several groups have demonstrated decreases in the retention times of polar analytes through the use of mobile phase modifiers [11–14]. The use of modifiers can be broadly placed into two classifications. First, when the amount of modifier added to the mobile phase is relatively small, its principle effect is to neutralize or hinder the sorbent active sites. The second case is when modifier is added to the mobile phase in high enough concentrations to enhance the solubility of the analyte in the mobile phase. According to *Board et al.*, the level of modifier necessary to achieve this latter effect is approximately greater than or equal to 1.0% [11]. At modifier levels below 1.0%, the blocking of sorbent active sites is probably the main effect facilitating the elution of polar analytes.

Many compounds have been studied as modifiers in packed column SFC. The problem with most modifiers is that they produce a significant response in the FID. When using an FID, this makes pressure or density programming unfeasible due to severe baseline rise. It is difficult to perform even isobaric work under these conditions as the small signal current produced by an eluting peak is obscured by the large background current produced by the modifier, leading to poor sensitivity and reproducibility. Therefore much of the modifier work that has been done to date has been performed using an ultra-violet (UV) absorption detector.

One polar modifier that does elicit a very small response to the FID is formic acid, which has successfully been used as a mobile phase modifier by several groups [8, 12]. Formic acid is now commonly used at the 0.3% level in CO_2 to hinder the active sites in packed columns. A separation performed with such a system is shown in **Figure 4**, where a series of silanol-terminated polysiloxanes have been eluted on a C_{18} packed column with pressure programming. To date, formic acid has been used as a modifier up to a level of 3.0% [15]. It should be pointed out that the formic acid used to modify the mobile phase must be free of impurities that respond to the

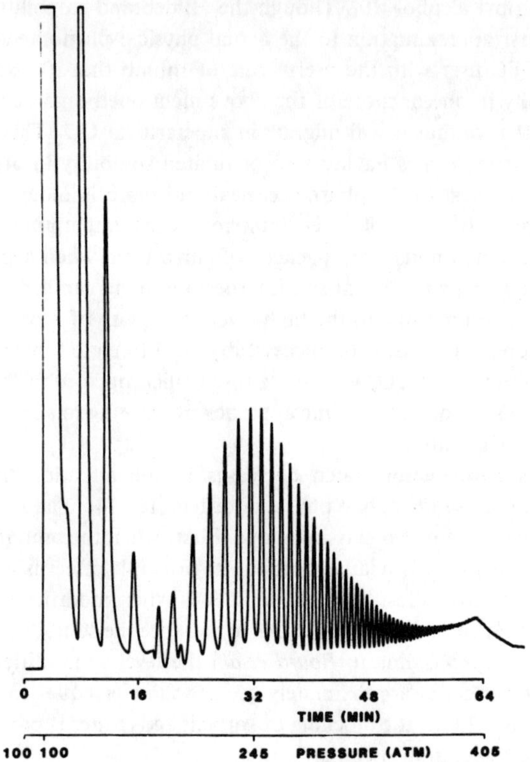

Figure 4
Packed column SFC separation of silanol-terminated polysiloxanes (M.W. 400–700). Column: octadecylsilane packed column, 5 μm particle size, 10 cm x 1 mm i.d.. Mobile phase: CO_2 with 0.3% formic acid. Conditions: 100 atm for 3 min, then programmed at 5 atm/min. T_{col}: 150 °C. Detector: FID.

FID, as these can cause significant baseline rise, especially at increasing modifier concentrations. More work is needed to develop methods for purification of formic acid for use as a modifier in SFC.

The fact that most modifiers are not compatible with the FID has been mentioned earlier. A UV detector may be used as an alternative, but is not nearly as universally applicable to organic compounds as is the FID. A mass selective detector (MSD) or a mass spectrometer (MS) are excellent near-universal detectors that are experiencing increased use in SFC [16–22]. Virtually any modifier can be used with these systems, provided that it is compatible with the SFC and detector hardware, and that the mass spectral pattern produced by the modifier does not interfere with the mass analysis region of the analytes being determined. As the molecular weights of typical SFC analytes are quite high (500 to ≈ 4000 amu), ths occurrence is unlikely. One disadvantage is that with present technology, any type of MS or MSD is much more expensive than in FID.

Rather than attempt to elute polar molecules by modifying CO_2, a different approach is to use a more polar mobile phase. The problem with this strategy is of a practical nature. The critical temperature of a compound tends to increase with increasing polarity. For example, the critical temperature of water is approximately 374.2 °C. Operating at higher temperatures negates one of the big potential advantages of SFC, that is to elute polar, thermally labile compounds at mild temperatures. One mobile phase that holds great promise is ammonia. (It should be noted that even though the critical temperature of ammonia is 132.4 °C, it is possible to operate at somewhat lower temperatures provided the two phase region of the phase diagram is avoided.) Ammonia does have the potential to elute polar compounds, but there are presently significant problems with respect to SFC hardware, column and detector compatibly, as well as a major safety issue. These must be addressed before it is practical to implement ammonia routinely.

2.4 Inertness

It has already been mentioned that packed columns currently used in SFC have a high degree of activity due to the presence of either unreacted bare silica sites, metal impurities, or residual polar species present as a result of the column manufacturing process. In reversed phase HPLC, the activity can be masked to a large extent by the presence of water, methanol, acetonitrile, ion-pairing reagents, or ion-suppression agents that are widely employed in the practice of the technique. In SFC, if pure CO_2 is used as the mobile phase, there are no polar species present to help neutralize active sites. This can result in severe peak tailing, irreversible adsorption, or higher elution densities for some polar analytes when compared to elution using bonded fused silica open-tubular columns.

There are three basic approaches to alleviating this problem. The first is to add a polar modifier to the mobile phase in the hope that it will neutralize the active sites. The use of modifiers in packed column SFC has already been discussed in some detail in Section 2.3.

A second approach to the problem is to choose a more polar mobile phase that may be more effective at masking active sites than is CO_2. This concept has also been discussed in Section 2.3 A third approach attempts to make the packed column itself more inert. Recently there has been increased effort in this area for both HPLC and SFC [23–26]. In particular, the work of *Schomburg et al.* [23] involving the crosslinking and bonding of GC-like polymeric phases to silica supports holds a great deal of promise for SFC. In this technique, a polymeric stationary phase is crosslinked and bonded to silica gel using a process similar to one used in the manufacture of modern wall-coated open-tubular fused silica GC columns [27]. The stationary phase thereby forms an "envelope" around the packing material in the hope that the column active sites will be effectively masked. Initial work has shown that at least some columns produced in this manner exhibit a significant decrease in activity. Caffeine is a polar compound that elutes with good peak shape by capillary SFC using unmodified CO_2. However, on a packed column, caffeine is either irreversibly adsorbed, or elutes with very poor shape, even when formic acid is added as a modifier. By far the best packed column chromatogram of caffeine produced to date was done by *Larry Taylor's* group using a cyanopropyl based polymeric phase bonded to silica [28]. These columns exhibit great potential, and will probably play a significant role in the further development of packed column SFC.

Because packed columns exhibit a relatively high degree of activity, it is reasonable to expect the elution density of a given solute to be higher with packed columns as opposed to open-tubular columns. The consequence of this situation would be decreased lower molecular weight range elution capability for packed columns. The elution densities for the hydrocarbon chromatograms in Figures 2 are presented graphically in **Figure 5**. Examination of these data reveal that the average hydrocarbon elution density for the packed column was slightly lower than for the open-tubular column. It should be noted that each of these separations were run under temperature, flow, and pressure programming conditions optimized for resolution and not minimum attainable elution densities. Also, no attempt was made to correct the data for differences in phase ratios or total phase present for the two columns. Nevertheless, these data demonstrate that for a practical separation of non-polar hydrocarbon solutes, the elution densities for the packed and open-tubular separations were approximately equal.

When the same experiment is run using more polar solutes, the results are quite different. The separation of an homologous series of silanol-terminated polysiloxanes by open-tubular and packed column SFC are depicted in **Figures 6** and 4, respectively. Note that even though the mobile phase used for the packed column separation contained 0.3% formic acid, the pressures required to elute the homologs is significantly higher than in the open-tubular case (about 60–80 atm for a given species). The fact that the packed column separation was performed at a temperature 10 degrees higher does not nearly account for this difference. These data clearly demonstrate the need for the development of more inert packed columns in SFC. It should also be noted that the packed column separation contains approximately 58 peaks (compared to around 40 for the open-tubular separation), revealing the

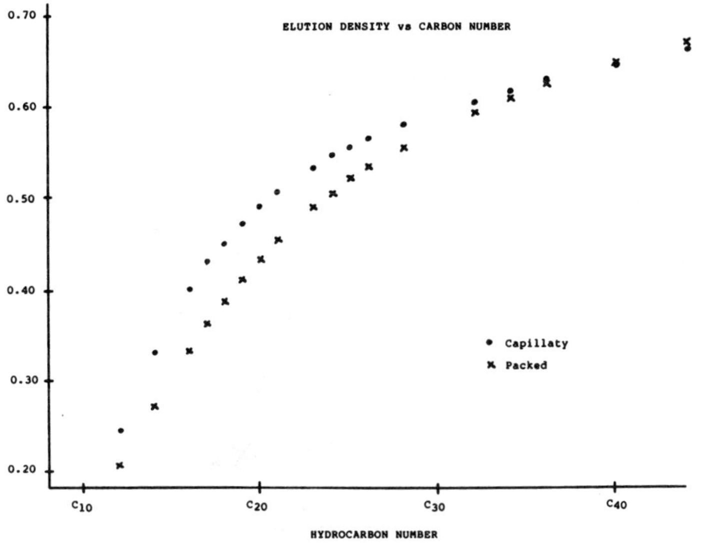

Figure 5
Plot of elution densities (g/ml) *vs.* solute carbon number for the packed and open-tubular hydrocarbon separations in Figure 2.

superior sensitivity attainable with packed columns and splitless injection as compared to open-tubular SFC with split injection.

2.5 Reproducibility

The area and retention time reproducibility obtainable in SFC is a topic that has not been widely addressed, though recently there have been some reports regarding the reproducibility obtainable with both packed and open-tubular SFC [29, 30]. Before addressing the subject of reproducibility, it is useful to briefly discuss injection techniques and hardware as applied to both packed and open-tubular SFC.

Generally, the types of injection valves used in SFC are identical to the internal loop rotor valves used in HPLC. The smallest commercially available internal loop rotor for this type of valve is 0.06 μl (60 nl). When using 50 μm and 100 μm i.d. open-tubular columns which have a maximum injection volume of 15 nl (when presently available conventional injection techniques are employed), only a fraction of the filled 60 nl sample loop can be injected onto the column. This is achieved

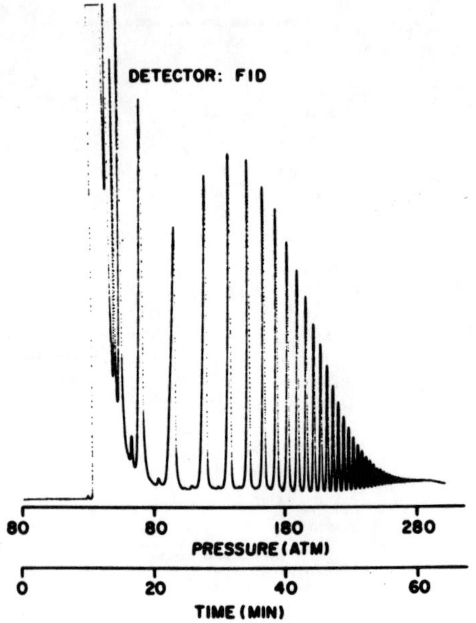

Figure 6
Capillary (open-tubular) SFC separation of silanol-terminated polysiloxanes (M.W. 400 – 700). Column: 10 m x 100 μm i. d. DB-5 fused silica capillary column, 0.25 μm film thickness. Mobile phase: CO_2. Conditions: 80 atm for 20 minutes, then programmed at 5 atm/min. Injection: 0.10 μl, split 10 : 1. T_{col}: 140 °C. Detector: FID.

either by split injection, or "timed partial loop" injection, which is also known as "digital valve interface" (DVI) injection. These injection techniques generally result in poorer reproducibility than splitless injection. Therefore reproducibility can be improved by eliminating the necessity for split or DVI injection. With narrow bore open-tubular columns, this has not as yet proven practical. However, when packed columns are employed, typical injection volumes range from 0.2 μl to 5 μl (or larger). Injection can therefore be made in a splitless mode, using the entire injection loop volume. **Table 2** lists the retention time and peak area reproducibility for a mixture of six vitamins separated on a cyano packed column in less than ten minutes. The relative standard deviations (r. s. d.'s) for retention time reproducibility are in the range of 0.30% to 1.18%, and the r. s. d.'s for area reproducibility vary from 1.30% to 3.39%. These values are quite good especially considering the polar, adsorptive nature of the analytes, and would generally be lower for less polar solutes. From this data, it is evident that there are no inherent properties or characteristics of packed column SFC that leads to unusually poor reproducibility.

Table 2
Packed column injection retention time and area reproducibility of six vitamins on a 10 cm x 1.0 mm i. d. packed cyano column (3 μm particle size); injection volume – 1 μl; CO_2 mobile phase, FID detection; (n = 9).

Packed column injection reproducibility				
	t_r (min)	Area (x 10^6)	t_r (min)	Area (x 10^6)
	Vitamin A Acetate		*Vitamin K_3*	
$\overline{X}$ =	2.01	15.31	3.51	1.292
S =	0.010	0.199	0.041	0.043
% r. s. d. =	0.50%	1.30%	1.18%	3.35%
	Vitamin K_1		*Vitamin E Acetate*	
$\overline{X}$ =	4.31	2.84	4.98	1.66
S =	0.047	0.087	0.040	0.044
% r. s. d. =	1.08%	3.07%	0.80%	2.62%
	Vitamin E		*Vitamin D_3*	
$\overline{X}$ =	6.52	4.228	9.89	4.13
S =	0.028	0.137	0.030	0.140
% r. s. d. =	0.43%	3.25%	0.30%	3.39%

2.6 Sensitivity

From statistics compiled in our applications laboratory, it is estimated that approximately 15–25% of samples received are of the type where the end goal of the chromatographic separation is resolution without regard to sensitivity. One may then ask: How useful is SFC for the analysis of samples containing relatively few components (*e. g.* 3–10) when one or more of the analytes of interest is present at "low" or "trace" levels?

To address this question, some conditions must first be established. Since the FID is the most common detector presently used in SFC, it is an appropriate starting point to begin a discussion of sensitivity. The response of the FID is proportional to the number of carbon atoms present in a molecule, and has been widely used in GC as a sensitive, universal detector [31]. To obtain the minimum signal/noise ratio that is required for the acquisition of reproducible quantitative data generally requires that one nanogram of analyte pass through the FID in a 20–30 second time period. As will be seen in a following section dealing with resolution, peak widths on this order can be attained with either packed or open-tubular SFC separations, while maintaining programming conditions that yield the separation of a large series of oligomers (*i. e.* resolution is not sacrificed). The remaining parameters that are important in determining the limit of detection are the background signal present, the maximum injection size possible, and the number of coulombs generated per gram of analyte in the FID under a specific set of operating conditions.

The absolute amount of noise observed can vary with such parameters as the purity of the mobile phase, the rate of pressure/density programming, the column temperature, and general cleanliness of the chromatographic system. The relative amount of noise in terms of a background signal that is observed for packed columns in SFC, is lower than in GC, where packed columns are generally regarded as having limited use in trace analysis. This is because many of the packed columns used in GC are of the type where the stationary phase is not immobilized by covalent bonding, but rather by adsorption onto a solid stationary phase support. These columns tend to exhibit much higher bleed rates than the bonded packed columns used in SFC, and therefore generate higher background currents, which ultimately results in decreased sensitivity.

The next consideration in the sensitivity equation is the total amount of current that the FID generates per gram of a particular analyte. A standard international system was originally developed so that the sensitivity of different FID's could be compared [32]. When this system was used to compare the coulombs per gram generated by an analyte by packed *vs.* open-tubular columns in SFC, they were found to be the same within experimental error (+10%) [33]. Therefore there is nothing inherent to packed column SFC that leads to poor FID ionization efficiency or ion collection.

The critical consideration in achieving lower detection limits then becomes the maximum volume that can be injected while maintaining acceptable peak width and analysis time. When a 1.0 mm i. d. packed column is used, it is feasible to inject volumes on the order of 0.2 to 5.0 μl without splitting. If an intermediate injection size of 3 μl is chosen, and the assumption is made that 1 ng of sample is needed for an acceptable *S/N* ratio, then the limit of detection that is theoretically achievable for packed column SFC with FID detection is approximately 0.25 ppm (w/w injected). Though this limit is theoretically attainable, a more practical range for present limits of detection lies between 1.0 and 5.0 ppm.

Figure 7 is a packed column separation of a standard mixture of artesunic acid and artemisinin. The concentrations of the two components in the sample solution were 3 ppm and 7.5 ppm respectively, with a 1 μl injection volume. Artemisinin is a sesquiterpene lactone endoperoxide isolated from the herb ginghao. Artesunic acid is a water-soluble ester derivative of artemisinin. These compounds are currently under investigation as antimalarial drugs, and therefore the clinical monitoring of these compounds at low levels is necessary.

As can be seen from Figure 7, at the low attenuation required to detect the analytes, the solvent peak extends 10 – 12 minutes into the run, and the baseline does

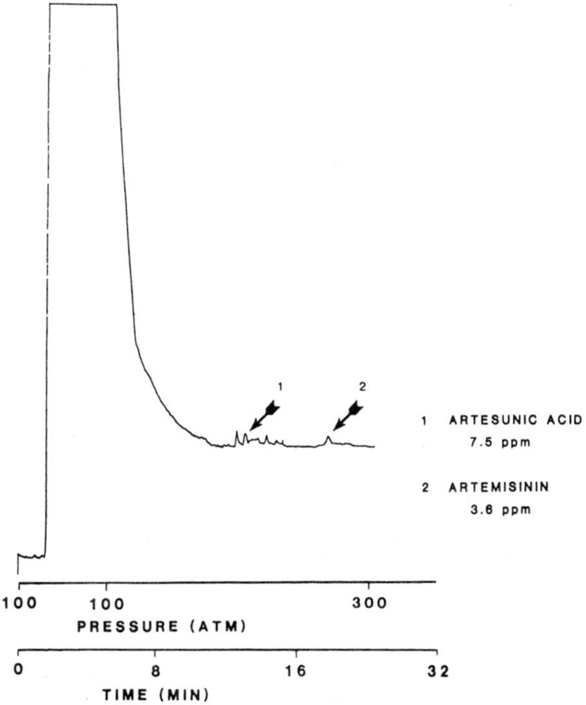

Figure 7
Packed column SFC separation of artesunic acid and artemisinin at low concentrations. Column: cyano packed column, 5 μm particle size, 10 cm x 1 mm i. d.. Mobile phase: CO_2 with 0.3% formic acid. Conditions: 100 atm for 5 min, then programmed to 300 atm at 5 atm/min. T_{col}: 80 °C. Detector: FID.

not return to its inital level. When the injection volume is increased to 5 μl in an attempt to improve detection limits, the solvent peak becomes much larger, and interferes with the analyte peaks and/or significantly lengthens the analysis time.

One solution to this situation involves the implementation of a modified version of a concentration technique that has been successfully used in both liquid chromatography (LC) and ion chromatography (IC). We refer to this modified system applied to SFC as "sensitivity enhancement". A diagram of this system is shown in **Figure 8**. It consists of a guard (concentrator) column, an analytical column, two six-port injection valves, and a purified helium or nitrogen source. Both the guard column and analytical column are located in thermally variable, heated

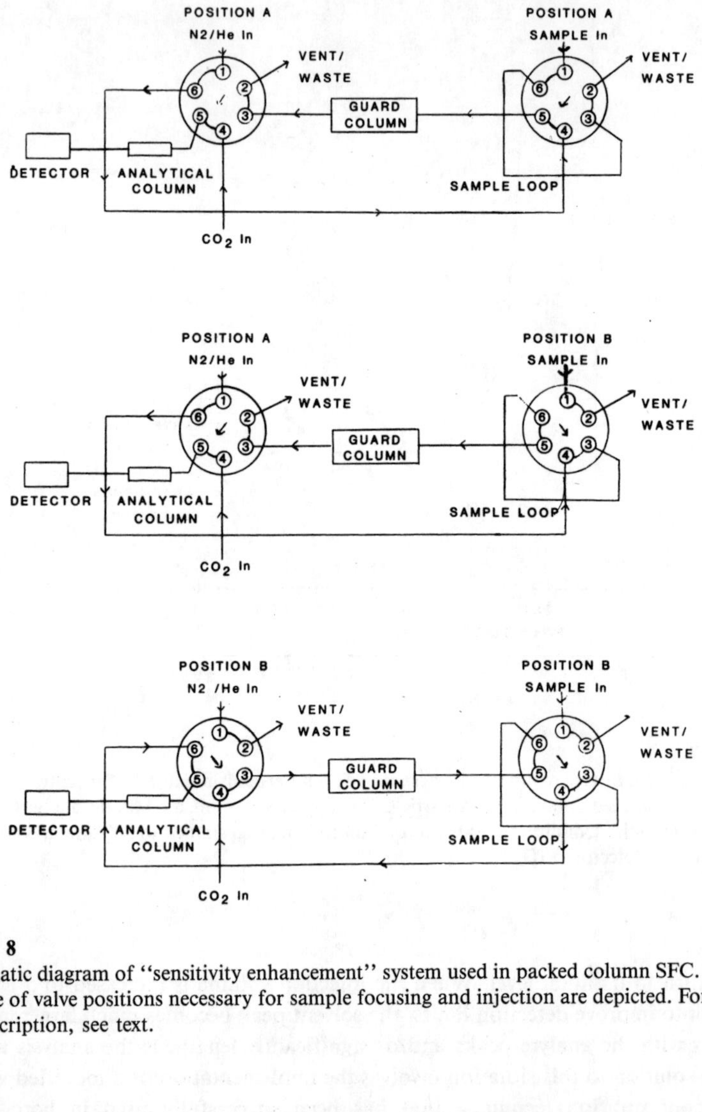

Figure 8
Schematic diagram of "sensitivity enhancement" system used in packed column SFC. The sequence of valve positions necessary for sample focusing and injection are depicted. For detailed description, see text.

compartments (ovens or blocks). In the uppermost position in Figure 8 the mobile phase moves from the pump directly to the analytical column, while the injection loop ($\approx$ 5 μl to 500 μl) is being filled and helium or nitrogen is flushed through the guard column. Next in the sequence, the sample loop is rotated from the load to inject position and the sample loop is flushed by the He/N_2 stream to the heated guard column. Under these GC-like conditions, the analytes, being of higher molecular weight, are adsorbed onto the front of the guard column, while the solvent passes through and out to vent. When this process is complete, the other six-port valve is rotated, directing mobile phase from the pump to the guard column, thus backflushing the sample onto the analytical column. The times necessary for each of the steps of this sequence will vary with injection volume, temperature, initial density, flow rate of He/N_2, *etc.*, but is generally in the range of 3–12 minutes total. This system allows larger volumes to be injected and therefore lowers detection limits, while eliminating the large solvent peak from the chromatogram. Injection volumes up to $\approx$ 100 μl have been used for initial work [8], and it should be feasible to extend this to at least 1.0 ml with a better understanding of the optimization of the parameters of the sensitivity enhancement system. This could lower future packed column SFC/FID detection limits to the 10–50 ppb range.

2.7 Resolution

For a useful discussion of resolution in SFC, there are two basic cases to consider. The first involves isobaric or constant density separations. In this mode, one can invoke the theoretical plate model derived from distillation theory. A theoretical plate occupies a length of column which in principle is known as the height-equivalent-to-a-theoretical-plate (HETP). The larger the number of theoretical plates for a given length of column, the narrower the peaks and the higher the efficiency. Therefore, in the absence of any pressure, density, temperature, or flow programming, the linear velocity can be optimized under a given set of chromatographic conditions to produce the most efficient separation. This applies to both packed column and open-tubular separations.

For packed columns, *Gouw* and *Jentoft* report that isobaric or constant density conditions result in an inefficient chromatographic process [34]. The authors explain this by describing the increase of partition ratios with pressure drop as a solute travels down the column and suggest that this situation in SFC is analogous to an inverse temperature program in GC. From these statements, the recommendation for isobaric work with packed columns would be to work under low pressure drop conditions (short columns and appropriate areas of the pressure/density isotherms) and take advantage of the ability of packed columns to generate large numbers of theoretical plates in short periods of time [30].

There has been one report in the literature that directly contradicts the above rationale. In 1982, *Gere*, *Board*, and *McManigill* reported a separation of anthracene, perylene, and coronene performed isobarically on a packed column [35]. Experimental conditions were chosen so that there was a very large pressure drop across the column. The column was a 3 μm particle size, 15 cm x 4.6 mm bonded octadecylsilane column. The resulting pressure drop was 123 bar, the inlet density being equal to 0.82 g/cm³, while the outlet density was 0.60 g/cm³. This is a significantly large pressure drop relative to normal operating conditions, and therefore presumably represents a worse case situation. The chromatogram obtained under these conditions is shown in **Figure 9**. The peaks are symmetrical and there is no significant difference in the reduced plate heights for a large range of k' values (H/d_p = 12.5, 9.8, and 8.2 for anthracene, perylene, and coronene respectively). The total number of plates calculated using coronene is approximately 6,100. From these data, a fundamental compromise between column efficiency and column pressure drop is not evident.

The second case is the situation where SFC is performed in conjunction with pressure, density, temperature, flow, or gradient programming. Though theory has been developed for HPLC with gradient elution, there has not been an analogous theory developed to explain the effects observed with programming in SFC. It is therefore useful to employ the concept of separation number, a simple practical means of comparing the "goodness" of two chromatograms such as the pressure programmed hydrocarbon separations in Figure 2. The larger the separation (Trennzahl) number for a given peak pair, the greater the number of peaks of similar width could theoretically be fit between the pair (while maintaining baseline resolution), and therefore the greater is the resolution.

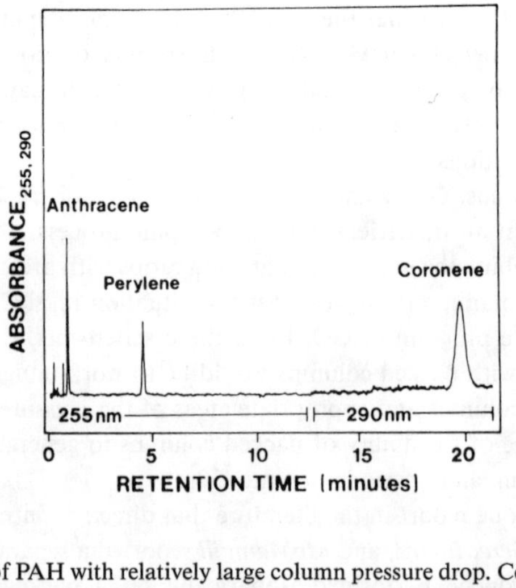

Figure 9
SFC chromatogram of PAH with relatively large column pressure drop. Column pressure drop: 123 bar. Column inlet pressure: 261 bar. Column outlet pressure 139 bar. Average column pressure, 200 bar; inlet density, 0.82 g/cm³; outlet density, 0.60 g/cm³; outlet density, 0.60 g/cm³; column temperature, 55 °C. (Reprinted with permission from reference [35].)

The Trennzahl numbers for the capillary and packed column hydrocarbon separations depicted in Figure 2 are presented graphically in **Figure 10**. Examination of this data reveals that for the earlier eluting peak pairs, the separation numbers for the open-tubular separation are significantly larger than for the packed column chromatogram. This difference decreases with increasing carbon number and converges at C_{28}. For hydrocarbon pairs above C_{28}, the separation numbers calculated from the packed column chromatogram are actually slightly higher than for the open-tubular case.

Why do packed columns compare so well to open-tubular columns in SFC? Though there is no single answer to this question, several explanations are plausible and each probably contributes to a greater or lesser degree. First, packed columns used in SFC are in the particle size range of 3 – 10 μm, and therefore yield higher efficiencies than the larger particle sizes used in packed GC columns. The second part of the explanation may lie in the small but finite pressure drop that exists across an SFC packed column, created by the small particle size packings typically employed. Though it has been claimed to be detrimental to chromatographic efficiency [3, 34], this pressure drop may actually be an asset in the presence of pressure or density programming. This concept is not new, and has been described previously

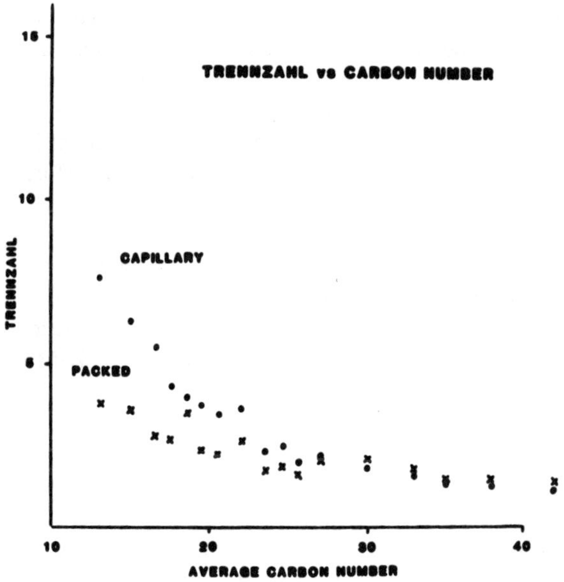

Figure 10
Plot of Trennzahl (separation) number *vs.* solute carbon number for the packed and capillary hydrocarbon separations in Figure 2.

by *Jeff Graham* and *L. B. Rogers* [36]. During pressure or density programming, solutes begin to migrate down the packed column but begin to precipitate as they enter lower density regions in the column. The solute remains there until the density in that region increases to the point where the solute again begins to migrate. This process occurs continuously down the length of the column, the leading edge of the solute band always being subjected to lower densities and pressures than the trailing edge of the band. This process serves to compress and refocus the solute, the end result being a very narrow chromatographic peak. The situation is analogous to that in packed column gas chromatography, where a decreasing temperature gradient is created across the column. This type of system has been described and given the name ''chromathermography'' by *Ohline* and *DeFord* [37]. These authors in fact report a band compression phenomenom for their GC experiments that may also exist in packed column SFC.

The third possible explanation is so simple that it might easily be overlooked. The analysis time limit chosen for the separation comparison in Figure 2 was ≈ 30 minutes. It is possible that this relatively short period of time, though a good practical limit for a routine analysis, may favor packed columns. In open-tubular columns, molecules must diffuse greater average distances to partition between the mobile and stationary phases than for packed columns. To achieve the open-tubular separation shown in Figure 2, it is quite possible that the combination of linear velocity, temperature, and pressure programming (though optimized for a 30 minute analysis time), does not nearly approach the ultimate resolving power of the open-tubular system that might be realized with longer analysis times (*e.g.* 1–3 h). On the other hand, the ultimate resolving power of packed columns may be closer to the optimized conditions for a 30 minute analysis time, and therefore resolution may not improve as much with longer analysis times.

Three possible explanations have been presented in an attempt to explain the data depicted in Figure 2. The magnitude of the relative contribution of each is presently unknown. Also, other as yet unidentified phenomena may be important. Present and future work will involve experimentation that will yield a better understanding of the important parameters affecting resolution for both packed and open-tubular columns in SFC.

2.8 Conclusions

Packed columns were used in nearly all of the early developmental work in SFC. Around 1981, research efforts began to focus on open-tubular SFC, perhaps driven at least in part by a similar trend occurring in GC. More recently, as SFC users begin to realize the limitations of open-tubular SFC, packed columns are receiving increased attention.

Perhaps the biggest myth surrounding the use of packed columns is the notion that the resolution obtained is far inferior as compared with open-tubular SFC. The separations depicted in Figures 2, 4, and 6 illustrate that under analysis time constraints of $\approx$ 1 h or less, the resolution attained under optimized operating conditions for each column type is very similar. Since peak widths generated with both types of columns are comparable, packed columns currently offer lower sample detection limits due to their higher volume capacity. Through the use of a dual six-port valve "sensitivity enhancement" system, it is quite possible that the detection limits attainable with packed column SFC/FID could reach the 10–50 ppb range.

The biggest disadvantage of currently available packed columns is their high level of activity (relative to open-tubular columns). This can result in irreversible solute adsorption, peak tailing, and increased solute elution densities. The problem is currently being addressed through the development of modified mobile phase systems, the use of mobile phases other than CO_2, and the development of more inert packed columns. The first commercially available crosslinked and bonded polymeric SFC packed columns have recently been introduced [38].

Because packed columns have many properties that are desirable for SFC, their further development is very worthwhile, and is being pursued by an increasing number of groups [39–50]. Packed columns will ultimately play an essential role in the commercialization and widespread acceptance of SFC.

References

[1] *E. Klesper, A. H. Corwin,* and *D. A. Turner,* J. Org. Chem. **27** (1962) 700–701.

[2] *C. M. White,* "SFaCts – A Comprehensive Bibliography of SFC and Related Articles", White Associates, Pittsburgh, PA. 1987.

[3] *M. Novotny, S. R. Springston, P. A. Peaden, J. C. Fjeldsted* and *M. L. Lee,* Anal. Chem. **53** (1981) 407A–414A.

[4] *F. Pacholec,* LC/GC Magazine **4** (1986) 432–441.

[5] *W. Jennings,* J. Chromatogr. Sci. **17** (1979) 637.

[6] Private communication, *Dr. Dick Henry,* October, 1987.

[7] *L. R. Snyder* and *J. J. Kirkland,* "Introduction to Modern Liquid Chromatography", 2nd Edition, p. 36, Wiley Interscience, New York (1979).

[8] *F. Pacholec, D. S. Boyer, A. C. Rosselli,* and *R. M. Ravey,* unpublished results, 1987.

[9] *J. Buchner* and *R. Willoaghby,* Paper #265, Rocky Mountain Analytical Symposium, Denver, CO, 1987.

[10] *L. M. Bowman Jr.,* Doctoral Dissertation entitled "Dense Gas Chromatographic Studies", University of Utah, Salt Lake City, 1976.

[11] *R. Board, D. McManigill, H. Weaver,* and *D. Gere,* presented at the 1982 Pittsburgh Conference on Analytical Chemistry.

[12] *J. M. Levy* and *W. M. Ritchey,* J. Chromatogr. Sci. **24** (1986) 242–248.

[13] *J. M. Levy* and *W. M. Ritchey,* HRC & CC **10** (1987) 493–496.

[14] *C. R. Yonker, D. G. McMinn, B. W. Wright,* and *R. D. Smith,* **396** (1987) 19–29.

[15] *F. Pacholec, C. M. White, D. S. Boyer,* and *R. M. Ravey,* presented at the 1987 Pittsburgh Conference, Atlantic City.
[16] *S. B. Hawthorne* and *D. J. Miller,* J. Chromatogr. **388** (1987) 397–409.
[17] *H. T. Kalinoski, H. R. Udseth, B. W. Wright,* and *R. D. Smith,* J. Chromatogr. **400** (1987) 307–316.
[18] *H. T. Kalinoski, H. R. Udseth, E. K. Chess,* and *R. D. Smith,* J. Chromatogr. **394** (1987) 3–14.
[19] *R. D. Smith, H. R. Udseth,* and *B. W. Wright,* J. Chromatogr. Sci. **23** (1985) 192–199.
[20] *J. B. Crowther* and *J. D. Henion,* Anal. Chem. **57** (1985) 2711–2716.
[21] *E. D. Lee* and *J. D. Henion,* HRC & CC **9** (1986) 172–174.
[22] *R. D. Smith* and *H. R. Udseth,* Anal. Chem. **55** (1983) 231–243.
[23] *U. Bien-Vogelsang, A. Deege, H. Figge, J. Kohler,* and *G. Schomburg,* Chromatographia **19** (1984) 170–179.
[24] *H. Figge, A. Deege, J. Kohler,* and *G. Schomburg,* J. Chromatogr. **351** (1986) 393–408.
[25] *G. Schomburg, A. Deege, J. Kohler,* and *U. Bien-Vogelsang,* J. Chromatogr. **282** (1983) 27–39.
[26] *G. Schomburg, J. Kohler, H. Figge, A. Deege,* and *U. Bien-Vogelsang,* Chromatographia **18** (1984) 265–274.
[27] *W. G. Jennings,* ''Comparisons of Fused Silica and Other Glass Columns in Gas Chromatography'', Dr. Alfred Hüthig Verlag, Heidelberg (1981).
[28] *Prof. Larry Taylor,* presented at SFC '87, Pittsburgh, PA.
[29] *Dr. Doug Later,* presented at SFC '87, Pittsburgh, PA.
[30] *S. K. Yarbro* and *D. R. Gere,* Chrom. Forum, April (1987) 49–55.
[31] *C. F. Poole* and *S. A. Schuette,* ''Contemporary Practice of Chromatography'', Elsevier, Amsterdam, 1984.
[32] International Organization of Legal Metrology, ''International Recommendation on Gas Chromatograph/Mass Spectrometer/Data System for Analysis of Organic Pollutants in Water'', 1984.
[33] *D. R. Gere,* unpublished results.
[34] *T. H. Gouw* and *R. E. Jentoft,* J. Chromatogr. **68** (1972) 303–323.
[35] *D. R. Gere, R. Board,* and *D. McMannigill,* Anal. Chem. **54** (1982) 736–740.
[36] *J. A. Graham* and *L. B. Rogers,* J. Chromatogr. Sci. **18** (1980) 75–84.
[37] *R. Wayne Ohline* and *D. D. DeFord,* Anal. Chem. **35** (1963) 227–234.
[38] Keystone Scientific, State College, PA.
[39] *H. E. Schwartz,* LC/GC Magazine **5** (1987) 14–22.
[40] *H. E. Schwartz, P. J. Barthel, S. E. Moring,* and *H. H. Lauer,* LC/GC Magazine **5** (1987) 490–497.
[41] *H. E. Schwartz* and *R. G. Brownlee,* J. Chromatogr. **353** (1986) 77–93.
[42] *P. J. Schoenmakers* and *F. C. C. J. G. Verhoeven,* J. Chromatography **352** (1986) 315–328.
[43] *P. J. Schoenmakers,* 8th International Symposium on Capillary GC, ''Supercritical Fluid Chromatography: Open Columns *vs.* Packed Columns'' **2** (1987) 942–954.
[44] *F. P. Schmitz* and *E. Klesper,* Polym. Commun. **24** (1983) 142–144.
[45] *F. P. Schmitz, H. Hilgers,* and *E. Klesper,* J. Chromatogr. **267** (1983) 267–275.
[46] *L. G. Randall,* Hewlett Packard Technical Paper #43–5953–1722 (1983) –16.
[47] *D. Ishii, Y. Hashimoto,* and *T. Takeuchi,* 8th International Symposium on Capillary GC, ''Micropacked Column Liquid and Supercritical Fluid Chromatography with Carbon Dioxide'' **2** (1987) 996–1004.
[48] *Y. Hirata* and *F. Nakata,* J. Chromatogr. **295** (1984) 315–322.

[49] *Y. Hirata, F. Nakata,* and *M. Kawasaki,* HRC & CC **9** (1986) 633–637.
[50] *M. Caude* and *R. Rosset,* Analusis **14** (1986) 310–311.

3 Solubility of Various Organic Modifiers in Liquid Carbon Dioxide

Karen L. Maguire and Robert B. Denyszyn
Scott Specialty Gases Plumsteadville, PA 18949, USA

Karen L. Maguire is a research chemist with the Research and Development department of Scott Specialty Gases, a division of Scott Environmental Technology, Inc. She received her B. S. in chemistry from the Delaware Valley College of Science and Agriculture, Doylestown, PA, in 1988.

Dr. Robert B. Denyszyn is technical director at Scott Specialty Gases where he directs engineering, research and development, and quality assurance activities. He received his B. S. in chemistry from Lynchburg College; Lynchburg, Virginia in 1971. He received his master degree in analytical chemistry under the direction of *Dr. A. Bandy* at Old Dominion University, Norfolk, Virginia in 1977. His thesis topic involved studies of naturally emitted hydrocarbons in the great dismal swamp. He continued developing gas chromatographic methods at the Research Triangle Institute from 1974 to 1978, and then joined Scott Specialty Gases as senior chemist for one of Scott's largest quality control labs. He finished his doctorate work at Drexel University studying the interaction of carbon monoxide on metal surfaces. He was recently appointed to a National Research Council panel which reviews the work of the National Bureau of Standards. He has published numerous articles on trace and high accuracy measurement of various gaseous species.

Summary:

The presence of modifiers in the carbon dioxide mobile phase can enhance analysis by supercritical fluid chromatography. However, it is important that the chromatographer know the optimal conditions under which to use these mixtures to ensure homogeneity. Until recently, there has been little data for the prediction of polar modifier solubility in carbon dioxide. The solubility behavior of three polar modifiers: methanol, acetonitrile, and chloroform in carbon dioxide is presented and interpreted for temperatures ranging from −4 to 50 °C and pressures from 15 to 170 atm.

The modifier and carbon dioxide were combined in a stainless steel extraction apparatus so that the system was saturated with modifier. Samples were injected onto an HP 5710A gas chromatograph (FID) containing a megabore column.

In the methanol/carbon dioxide mixture, pressure and temperature had little effect on solubility below 90 atm. However, the impact of both parameters is significant above 90 atm. For the mixture to be homogeneous, the methanol concentration must be kept below 12 mole %. The solubility of acetonitrile in carbon dioxide was more dependent on temperature than on pressure. There was a small increase in solubility with increased pressure and temperature. To ensure homogeneity, acetonitrile concentration must be kept at or below 1.6 mole %. Both temperature and pressure had a definite influence on the solubility of chloroform in carbon dioxide. The concentration of modifier increased steadily as temperature and pressure rose. A chloroform concentration of 1.4 mole % or less is necessary for a homogeneous mixture.

Concentrate:

The importance of temperature and pressure effects on the homogeneity of a carbon dioxide mobile phase with a polar modifier, such as methanol, acetonitrile, or chloroform is illustrated. In order to maintain a homogeneous mixture, the chromatographer must use concentrations of modifier at or below 12 mole % for methanol, 1.6% for acetonitrile, and 1.4% for chloroform, or must carefully regulate pressure and temperature to prevent formation of a two-phase system.

3.1 Introduction

3.1.1 Review of Past Solubility Research

The solubility of organic compounds in carbon dioxide has been a subject of great interest to researchers for many years. The following summarizes some of the important work done in this area.

The first extensive work on solubility in liquid carbon dioxide was done by *Alfred W. Francis* in 1954. In this work, he discussed the mutual solubilities of carbon dioxide with 261 other substances. He showed 464 triangular graphs of ternary systems involving liquid carbon dioxide about room temperature. Nearly half of the substances tested were miscible with carbon dioxide and almost every graph showed its homogenizing effect [1].

Much of the research in the years following *Francis'* work deals with binary mixtures of solids or liquids with carbon dioxide at subcritical and supercritical conditions.

Two research groups who have studied the solubility of liquids in carbon dioxide involved *John M. Prausnitz, A. D. King, Jr.,* and their co-workers. In 1959, *Prausnitz* and *Benson* investigated the solubility of CCl_4, isooctane, toluene, and *n*-decane in hydrogen, nitrogen, and carbon dioxide at 50 and 75 °C, and at pressures of 20–90 atmospheres. They determined from this work that the vapor phase quickly departs from ideality as pressure increases, especially at pressures above 10 atmospheres [2].

In the early 1970's, *A. D. King* and his associates authored a series of four papers on the solubility of alcohols in compressed gases. They investigated the behavior of water, methanol, ethanol, and butanol at pressures up to 60 atmospheres and at temperatures up to 100 °C. From the data in each paper, the second cross-virial coefficients were evaluated and compared to those of homomorphic compounds. In each case, the coefficients were found to be more negative for the alcohols with carbon dioxide than for the homomorphs with carbon dioxide. These results were interpreted as indicating a reversible chemical association between the alcohol and carbon dioxide in the gas phase [3–6]. In the fourth paper, diethyl ether was also studied and was determined to have a charge transfer association with carbon dioxide which led the authors to conclude that the alcohol/carbon dioxide association was probably similar [6].

In the 1980's, *Prausnitz* and associates presented two papers on polar compound solubility in various compressed gases. One discussed the solubility of methanol from −45 to 10 °C and from 17 to 62 bars and used this data to calculate solvent losses in the gas absorption process [7]. The other investigated the behavior of acetone and isopropyl ether in nitrogen, methane, and carbon dioxide from −50 to

50 °C and from 17 to 63 bars. The data was used to determine the second cross-virial coefficients which were then compared to those calculated by the square well potential and the Hayden-O'Connell correlation. The authors determined that the square well and Hayden-O'Connell methods do not work for carbon dioxide for two possible reasons. There may be a strong interaction between carbon dioxide and acetone (isopropyl ether) or the calculated coefficients are in error due to uncertainties in the Henrie's constants [8].

Other groups have researched the solubility of solids in compressed gases since supercritical fluid extraction has been found useful in separating solids where other extraction techniques have been unsuccessful.

In 1970, *J. Calvin Giddings,* and associates published a paper on the solubility of several solids of varying molecular weights in compressed carbon dioxide at 40 °C and from 270 to 1900 atmospheres. They measured the solubilities on a high pressure gas chromatograph and found maxima in all curves between 300 and 2500 atmospheres. In a comparison of theory to experiment, they found qualitative agreement and some quantitative predictability [9].

Ten years later, *Ziad Alwani* investigated the solubility of non-volatile biochemical substances in compressed gases using tocopheryl acetate in carbon dioxide as an experimental model. He used a new optoelectrical method for the measurements which enabled him to determine relatively low solubilities. The solubilities were determined as a function of pressure at a given temperature. *Alwani* found that the solubility of tocopheryl acetate increases with pressure in liquid and supercritical carbon dioxide and changes only slightly when carbon dioxide is heated from subcritical to supercritical temperatures at a constant pressure [10].

In 1981, *Kurnik, Holla,* and *Reid* studied the solubility of five different solids in supercritical ethylene and carbon dioxide using a one pass flow system. Measurements were made at temperatures ranging from 35 to 65 °C and pressures from 80 to 280 bars. They found that the ideal gas assumption holds only for low pressures and concentration is greatly underestimated at high pressures [11].

As part of the book "Chemical Engineering at Supercritical Fluid Conditions" in 1983, *M. E. Paulaitis* and associates published their work on the solubility of solids in supercritical fluids at high pressures. Biphenyl and naphthalene were studied in ethylene and carbon dioxide. Pressure/temperature projections were constructed for each. The solid-liquid-gas equilibrium lines were interpreted using a model based on the Peng-Robinson equation and compared to other experimental results. Agreement between the two showed that the method is reliable [12].

Most recently, in 1986, a group of researchers, *Dandge, Heller,* and *Wilson,* published an extensive study of both solid and liquid solubility in carbon dioxide that is a supplement to the work of *Francis*. In this paper, they identify structural features that influence the solubility of different groups in carbon dioxide and give examples of these groups. The substances included are hydrocarbons, alcohols, aldehydes, ethers, phenols, esters, amines, and nitro compounds. The results of this study can be used to predict qualitatively the solubility of many compounds in carbon dioxide [13].

3.1.2 The Purpose of Using Polar Modifiers in Liquid Carbon Dioxide

After reviewing the available literature, it is evident that most of the solubility research has been directed toward chemical engineering applications. However, chromatographers are also in need of solubility data. This data must cover wider ranges than presently reported in the literature. The solubility of polar modifiers in the carbon dioxide mobile phase is especially important for supercritical fluid chromatography (SFC) applications.

Existing literature shows that polar modifiers can be beneficial to the chromatographer for three important reasons. The first is that they enhance the solvating power of carbon dioxide with respect to certain substances, especially solids and more polar analytes. This phenomenon is shown in a paper by *Dobbs, Wong, Lahiere,* and *Johnston* in 1987. In this study, they investigated the solubility of several solids, such as benzoic acid and 2-aminobenzoic acid in carbon dioxide doped with varying amounts of acetone or methanol. The results indicated that the solubility of these solids was increased significantly when small amounts of the solvents were added. They also determined that the co-solvent/carbon dioxide mixtures can be very selective for certain solutes because of specific intermolecular interactions [4].

The second reason is that modifiers can alter retention characteristics, selectivity, and peak shape during SFC. There have been several papers published which discuss these effects. One of these was authored by *Blilie* and *Greibrokk* in 1985. In their study, they found that modifiers added to carbon dioxide resulted in shorter retention times and better peak shapes of polycyclic aromatic hydrocarbons, polystyrenes, and some other compounds [15]. Two others were written by *J. M. Levy* and *W. M. Ritchey* in 1985 and 1986. Both investigated the use of modifiers such as methanol, propanol, and acetonitrile in the SFC of several types of aromatic hydrocarbons. Using different modifiers they were able to obtain changes in retention and selectivity which showed that the use of modifiers could be advantageous for separation of complex mixtures [16, 17].

Another paper, authored by *D. Gere* and co-workers studied the effects of modifiers using various stationary phases, compounds, and modifiers. Their results showed that in changing the modifier concentration from 0 to 1% in the carbon dioxide mobile phase, retention could be altered more than 50 times. The magnitude of the effect depends upon the polarity and hydrogen bonding capabilities of the modifier. The retention alteration was shown to be true for polar modifiers only.

Nonpolar modifiers were observed to have no effect [18]. Additional work by *Gere* continued to show the value of polar modifiers, but also showed a significant impact on speed of analyses [19–23]. This third and final benefit is significant since chromatographers are always looking for methods which can maintain resolution and increase the speed of analyses.

However, at this time, very little data exists regarding the solubility of these polar modifiers in carbon dioxide over wide temperature and pressure ranges. Without this information, chromatographers cannot be certain of the homogeneity of the carbon dioxide/modifier system. This information is pertinent in view that such mixtures can be exposed to varied temperature/pressure regimes.

In summary, it is safe to say that little solubility data is available for binary polar modifiers at varying temperatures and pressures. Of the literature searched, only two sources were found for polar liquids at other temperatures and pressures. In "Chemical Engineering at Supercritical Fluid Conditions", a chapter by *Kuk* and *Montagna* investigated ethanol in carbon dioxide at 40 °C and 75 to 200 atmospheres and isopropanol in carbon dioxide at high pressures [24]. The other by *Panageotopoulos* and *Reid,* was in "Supercritical Fluids: Chemical and Engineering Principles and Applications". It discussed the solubility of acetone and ethanol in carbon dioxide at 30 to 60 °C and at 20 to 150 bars [25].

Even with these two studies, the issue of modifier solubility in carbon dioxide has barely been touched, especially for compounds such as methanol, acetonitrile, and chloroform. Therefore, to more accurately define the maximum solubility of these modifiers in carbon dioxide, a study was initiated. The temperatures to be evaluated ranged from −4 to 50 °C, and pressures ranged from 15 to 170 atmospheres.

3.2 Determination of Modifier Solubility

3.2.1 Experimental System

A stainless steel extraction apparatus was used to contain the carbon dioxide and modifier being studied (**Figure 1**). The apparatus consisted of a 300 ml high pressure sampling cylinder with a dip-tube extending approximately half the length of the cylinder. Three valves were connected to the cylinder: a carbon dioxide inlet valve, a sample outlet valve, and a liquid outlet valve. The carbon dioxide inlet valve was connected to a 300 cu. ft. cylinder of carbon dioxide with helium headspace and a dip-tube which was used to pressurize the sample cylinder with liquid carbon dioxide. The sample outlet valve was used to obtain the modifier/carbon dioxide mixture in equilibrium for analysis and the liquid outlet valve was used to withdraw the liquid modifier from the bottom of the cylinder.

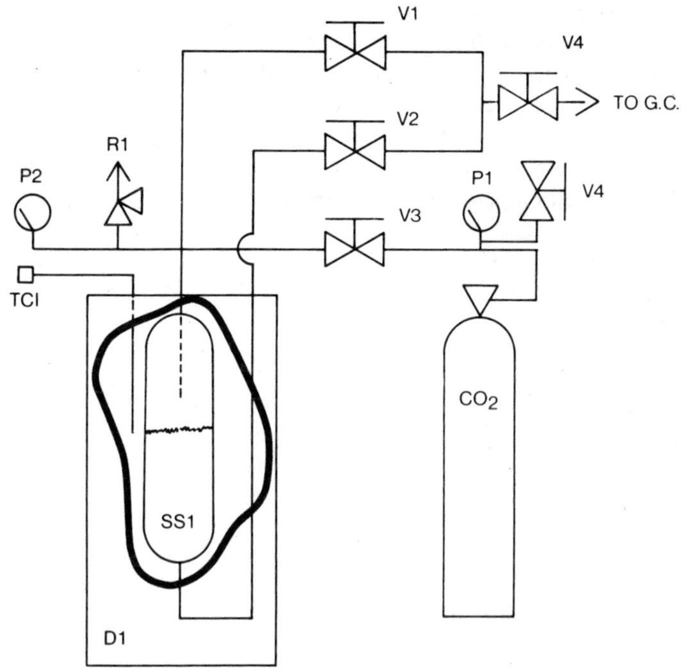

V1 Sample Withdrawl
V2 Liquid Withdrawal
V3 CO_2 Inlet
V4 Metering Valve
P1 CO_2 Pressure
P2 System Pressure

R1 Relief Valve
SS1 Sample Cylinder (w/dip tube) (300ml)
TCI Thermocouple
D1 Dewar Flask

Figure 1
Stainless steel extraction apparatus.

Connected to the sample outlet valve was a needle valve with 1/16″ stainless steel tubing leading to the injection port of the gas chromatograph. This line was heated above the boiling point of the modifier to prevent condensation.

Prior to filling, the sample cylinder was rinsed with a low boiling solvent, dried with nitrogen, and evacuated under a vacuum to 50 μm Hg. The evacuated cylinder was injected with 150 ml of modifier. The exact volume of modifier was determined by weighing the extraction apparatus before and after addition of modifier. All modifiers used were high purity solvents. After addition of the modifier, the extraction apparatus was immersed in an ethylene glycol/water bath that was kept at a constant temperature. The temperature was regulated by a glass immersion heater or liquid nitrogen and an electric stirrer. Once at the required temperature, the apparatus was pressurized using the cylinder of carbon dioxide.

All analysis were performed on a Hewlett-Packard 5710A gas chromatograph with a flame ionization detector. The following instrument settings were used:

Injection port	–	150 °C
Oven	–	80 °C
Detector	–	250 °C
Range	–	1000
Nitrogen carrier	–	2 cc/min
Column	–	J & W Megabore DB wax 30 meters long 1 μm film thickness

3.2.2 Experimental Measurements

Before analyzing a modifier/carbon dioxide mixture, a calibration curve for the modifier was constructed. Six standards between 1000 and 15,000 ppm of modifier in nitrogen balance gas were prepared in 5 liter glass flasks. A 1 cc sample loop was used for the analysis. A linear calibration curve was obtained for each modifier; area *vs.* concentration **(Figures 2** and **3)**.

Each day before and after the analysis of the modifier/carbon dioxide mixtures, a 10% mole/mole methane laboratory standard was analyzed to monitor the stability of the instrument response using the same sample loop used for the mixture (0.25 cc). Immediately before analyzing the mixtures, the extraction apparatus was inverted several times to ensure proper mixing. After equilibrating for 15 min, the sampling valve was opened slightly to allow a small amount of the mixture to enter the heated area between it and the needle valve. After closing the sampling valve, the needle valve was slowly opened to produce a steady flow through the sample loop. After the flow through the loop had slowed considerably, the needle valve was closed. After 20 s, the sample was injected onto the column. The injection valve was returned to the load position after 10 s. Within two minutes, a peak was eluted for the modifier. Several injections were made to obtain consistent results for each pressure at constant temperature and each range of pressures at constant temperatures were rerun several times to ensure reproducibility of the entire curve.

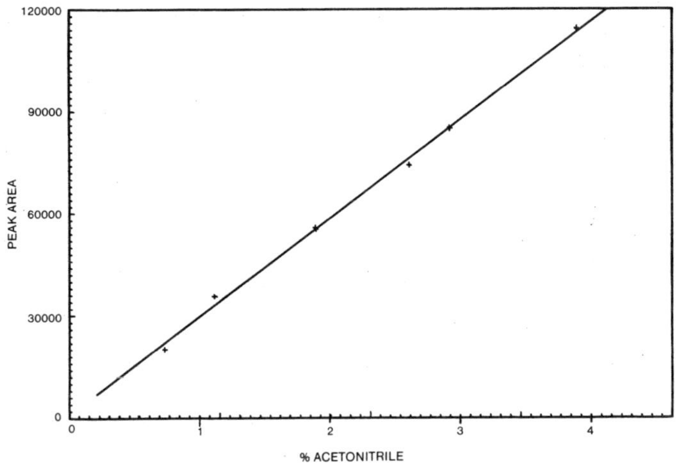

Figure 2
Calibration curve for acetonitrile in CO_2.

3.3 Results and Discussion

Three modifiers (methanol, acetonitrile, and chloroform) were analyzed in carbon dioxide for their solubility over a temperature range of −4 to 50 °C and a pressure range of 15 to 170 atmospheres. Curves for the data were obtained using least squares curve fits. Concentrations of modifiers were quantitated using the calibration curves generated.

3.3.1 Methanol/Carbon Dioxide Mixture (Figure 4)

The results of the analysis of methanol in carbon dioxide show that temperature and pressure have little effect on solubility when the pressure is below 90 atmospheres. However, the effects of both increase significantly at pressures above 90 atmospheres. In order to maintain a homogeneous mixture between −4 and 29 °C, the concentration of methanol as a modifier must be kept at or below 12 mole %.

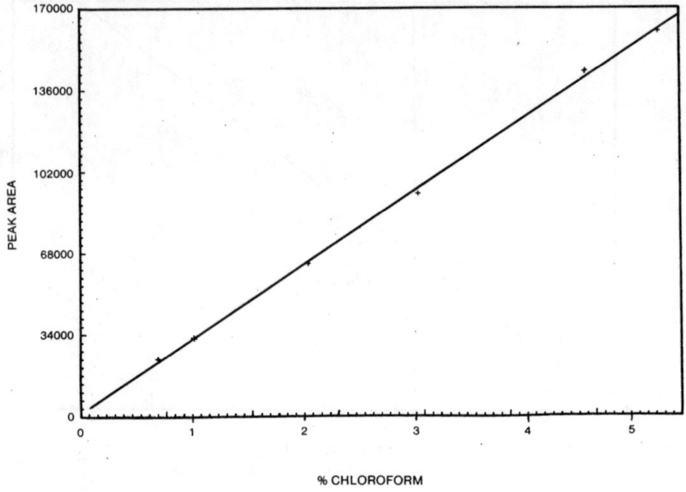

Figure 3
Calibration curve for chloroform in CO_2.

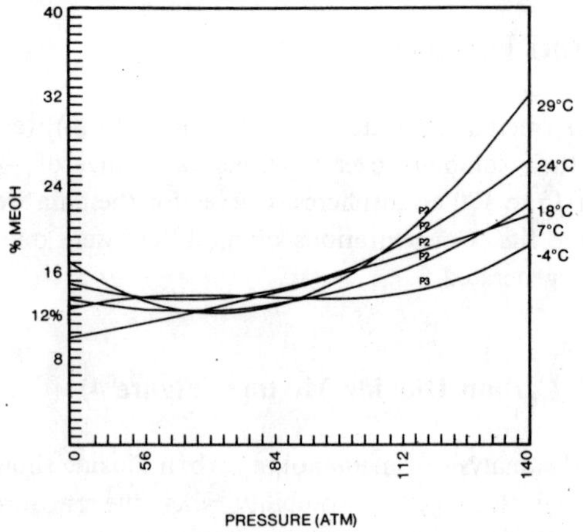

Figure 4
Methanol/CO_2 mixture.

3.3.2 Acetonitrile/Carbon Dioxide Mixture (Figure 5)

In the acetonitrile/carbon dioxide mixture, the solubility had very little pressure dependence but a small temperature effect was observed. At each pressure analyzed, solubility increased slightly with increasing temperature. To ensure homogeneity of the mixture at pressures up to 170 atmospheres and temperatures between 0 and 50 °C, the concentration of acetonitrile in the mixture should be kept below 1.6 mole %.

3.3.3 Chloroform/Carbon Dioxide Mixture (Figure 6)

For the mixture of chloroform in carbon dioxide, both temperature and pressure had a small effect on solubility. An increase in temperature or pressure resulted in an increased solubility of chloroform. In the pressure region of about 45 atmospheres, temperature seems to have no effect from 30 to 50 °C, but at 10 °C, the solubility is decreased. To maintain a homogeneous system at or above 30 °C in this mixture, the concentration of chloroform should be no higher than 1.8 mole %. However, if temperatures below 30 °C are to be used, the chloroform concentration should be held below 1.4 mole %.

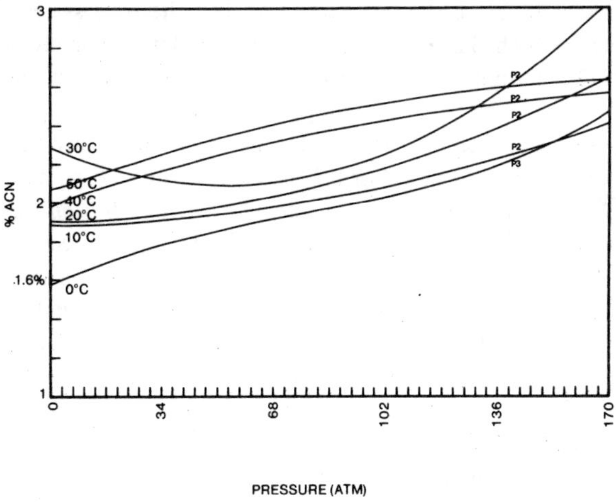

Figure 5
Acetonitrile/CO_2 mixture.

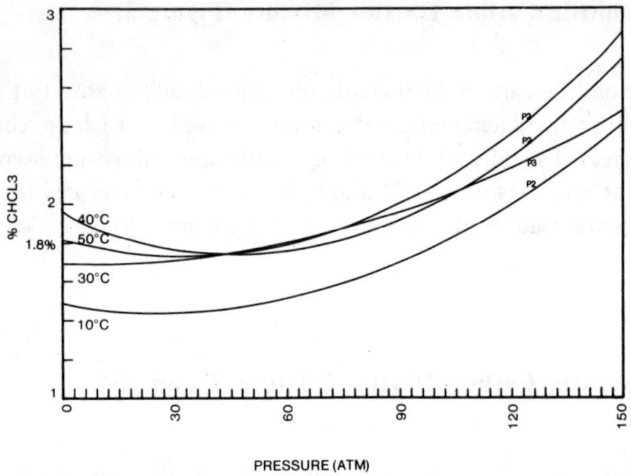

Figure 6
Chloroform/CO_2 mixture.

3.3.4 Comparison of Results to Literature

For the three systems investigated, it is difficult to effectively compare the data to previous research. From the literature available, the only system of the three which has been researched in depth is that of methanol in carbon dioxide. The work of *King* in 1972 determined that at low pressures, methanol is miscible in carbon dioxide to a much greater extent than in gases such as nitrogen, argon, or methane [4]. In 1980, *Prausnitz* and associates published data which gave the solubility of methanol in carbon dioxide as 0.2 or 0.3 mole % at –10 to 0 °C and about 17 atmospheres [7]. Another paper, by *Dandge, Heller,* and *Wilson* describes lower alcohols as being miscible in carbon dioxide at room temperature and 2500 psi [13]. Finally, in *Francis'* paper on ternary systems, methanol, acetonitrile, and chloroform were studied. According to *Francis*, all three are mutually miscible with liquid carbon dioxide. However, his data was obtained at room temperature only and the systems were not saturated with the liquid. It is also very difficult to make comparisons between his data and the data obtained in this study. *Francis'* phase diagrams all involved ternary systems. Therefore, the solubility of the modifier (methanol, acetonitrile, or chloroform) varied depending on the effects of the third component. For example, when mixed with carbon dioxide and kerosene (as in phase diagram A9), methanol is shown to be about 70% soluble. But, when mixed with carbon dioxide and lithium chloride (phase diagram A48), the solubility is closer to 20% and more like the results of the present study [1]. In view of the previous data, the only reasonable comparison that can be made is that both the previous literature and the present study agree that methanol is highly soluble in carbon dioxide and at room temperature, chloroform and acetonitrile are at least to some extent soluble in carbon dioxide.

3.4 Conclusion

The solubilities of methanol, acetonitrile, and chloroform in liquid carbon dioxide have been determined for temperatures ranging from −4 to 50 °C and pressures from 15 to 170 atmospheres. Comparison of the results with previous work is difficult due to the lack of data on the solubility of these compounds. However, it can be concluded that use of these modifiers in the carbon dioxide mobile phase at the maximum concentrations stated in the results section of this chapter should be avoided due to the possibility of forming a two-phase system. Further research into the area of modifier solubility is necessary in order to obtain information about other modifier/carbon dioxide systems since there are no models that can accurately predict solubility behavior of polar modifiers in general.

References

[1] *A. W. Francis,* J. Phys. Chem. **58** (1954) 1099–1114.
[2] *J. M. Prausnitz* and *P. R. Benson,* A. I. Ch. E. Journal **5** (1959) 161–164.
[3] *C. R. Coan* and *A. D. King Jr.,* J. Am. Chem. Soc. **93** (1971) 1857–1862.
[4] *B. Hemmaplardh* and *A. D. King Jr.,* J. Phys. Chem. **76** (1972) 2170–2175.
[5] *S. K. Gupta, R. D. Lesslie,* and *A. D. King Jr.,* J. Phys. Chem. **77** (1973) 2011–2015.
[6] *R. Massoudi* and *A. D. King Jr.,* J. Phys. Chem. **77** (1973) 2016–2018.
[7] *H. Lazalde-Crabtree, G. J. F. Breedweld,* and *J. M. Prausnitz,* A. I. Ch. E. Journal **26** (1980) 462–470.
[8] *P. J. Hicks Jr.* and *J. M. Prausnitz,* J. Chem. Engng. Data **26** (1981) 74–80.
[9] *J. J. Czubryt, M. N. Myers,* and *J. C. Giddings,* J. Phys. Chem. **74** (1970) 4260–4266.
[10] *Ziad Alwani,* Angew. Chem. Int. Ed. Engl. **19** (1980) pp. 623–623.
[11] *R. T. Kurnick, S. J. Holla,* and *R. C. Reid,* J. Chem. Engng. Data **26** (1981) 47–51.
[12] *M. E. Paulaitis, M. A. McHugh,* and *C. P. Chai,* in *M. E. Paulaitis, G. M. L. Penninger, R. D. Gray, Jr.,* and *P. Davidson,* eds., Chemical Engineering at Supercritical Fluid Conditions, Ann Arbor, Michigan, Ann Arbor Science (1983) pp. 139–158.
[13] *D. K. Dandge, J. P. Heller,* and *D. V. Wilson,* "Structure Solubility Conditions: Organic Compounds and Dense Carbon Dioxide Binary Systems", New Mexico Petroleum Recovery Research Center, New Mexico Institute of Mining and Technology (1986).
[14] *J. M. Dobbs, J. M. Wong, R. J. Lahiere,* and *K. P. Johnston,* Ind. Engng. Chem. Res. **26** (1987) pp. 56–65.
[15] *A. L. Blilie* and *T. Greibrokk,* Anal. Chem. **57** (1985) 2239–2242.

[16] *J. M. Levy* and *W. M. Ritchey,* HRC & CC **8** (1985) 503–509.
[17] *J. M. Levy* and *W. M. Ritchey,* J. Chromatogr. Sci. **24** (1986) pp. 242–248.
[18] *R. Board, D. McManigill, H. Weaver, D. Gere.,* "The Use of Modifiers in Supercritical Fluid Chromatography with Carbon Dioxide", presented at 1982 Pittsburgh Conference on Analytical Chemistry and Applied Spectroscopy.
[19] *D. Gere,* Hewlett Packard, Application Note AN 800–2, 1983.
[20] *D. Gere,* Hewlett Packard, Application Note AN 800–3, 1983.
[21] *D. Gere,* Hewlett Packard, Application Note AN 800–1, 1983.
[22] *D. Gere,* Hewlett Packard, Application Note AN 800–5, 1983.
[23] *D. Gere, T. J. Stark,* and *T. N. Tweeten,* Hewlett Packard, Application Note AN 800–4, 1983.
[24] *M. S. Kuk* and *J. C. Montagna,* in *M. E. Paulaitis, J. M. L. Penninger, R. D. Gray, Jr.,* and *P. Davidson,* eds., Chemical Engineering at Supercritical Fluid Conditions, Ann Arbor, Michigan, Ann Arbor Science (1983) pp. 101–111.
[25] *A. Z. Panagiotopoulos,* and *R. C. Reid,* in *T. G. Squires* and *M. E. Paulaitis,* eds., "Supercritical Fluids: Chemical and Engineering Principles and Applications", Washington, D. C., American Chemical Society, (1987) pp. 115–129.

4 Supercritical Fluid Chromatography/ Mass Spectrometry

Kent J. Voorhees, Steven D. Zaugg, and Stephan J. DeLuca
Department of Chemistry and Geochemistry, Colorado School of Mines, Golden, CO 80401, USA

Dr. Kent J. Voorhees is a professor of chemistry at the Colorado School of Mines. He received his Ph. D. in organic chemistry from Utah State University and has held post doctoral positions at Michigan State University and the University of Utah. He was director of the Analytical Chemistry section of the Flammability Research Center at the University of Utah and has been at the Colorado School of Mines since 1979. His major research interests have been in the areas of pyrolysis-mass spectrometry, chemometrics, and SFC/MS. He is the author of over 50 scientific publications and two patents in these areas. He has also edited a book on analytical pyrolysis. *Dr. Voorhees* is a member of the American Chemical Society and the American Society of Mass Spectrometry and serves on the Editorial Advisory Board for the Journal of Analytical and Applied Pyrolysis.

Dr. Steven D. Zaugg is a research scientist with the Water Quality Laboratory of the U. S. Geological Survey. He received his Ph. D. in analytical chemistry from Brigham Young University and spent two years as a post doctoral associate at the Colorado School of Mines with *Dr. Voorhees*. His research interests have been primarily in the areas of SFC/MS. *Dr. Zaugg* is a member of the American Chemical Society.

Dr. Stephan J. DeLuca is presently the Director of Research and Development for Quadraspec of Pittsburgh, PA. He received his Ph. D. from the Colorado School of Mines and was a post doctoral associate at the University of Colorado Medical Center. His research interest are SFC/MS, pyrolysis mass spectrometry, and biomedical instrumentation development. *Dr. Deluca* is a member of the American Chemical Society.

Summary:

Interfacing SFC with mass spectrometry (SFC/MS) offers a new dimension of selectivity and sensitivity, allows for broader analytical analysis, and provides identification information. Different applications, including thermally labile compounds, nearly intractable compounds, and highly polar molecules, have been published. The use of simple direct fluid injection (DFI) interfaces is possible for capillary SFC which has flow rates compatible with most conventional differentially pumped MS systems. The interfaces for chemical ionization (CI) SFC/MS are, for the most part, straightforward and have been used on an almost routine basis.

The application of electron ionization (EI) with SFC has been less successful. Poor sensitivity, solvent clustering, and higher than optimal source pressures have been reported. The approaches for overcoming the sensitivity problem including FT-mass spectrometry and novel interface designs are discussed. Detection levels approaching one ng have been reported using a two-stage heater probe. The FT-MS, because of its ion trapping capabilities, allows simple DFI probes to be used while maintaining acceptable detection levels. The major developments of SFC/MS are presented and the future prospects of commercial interfaces for SFC/MS are also discussed.

4.1 Introduction

The benefits of supercritical fluid chromatography (SFC) have been extensively discussed [1, 2]. The properties of the fluid such as solvent power, diffusivity, viscosity, and new commercial interests, makes SFC an exciting new entry into the chromatography arena. As is the case with most developing techniques, there are limitations associated with SFC which have to be solved in order for the approach to reach its full potential. One of these drawbacks has been the lack of a universal detector since FID is only compatible with a few fluids. There are a number of analytical instruments which could be considered for this application; however, mass spectrometry because of its application with gas chromatography is in the strongest position to fulfill the demanding requirements of a universal detector. Mass spectrometry has no optical absorbance requirements, is only minimally affected by the type of fluid, has the necessary sensitivity, and provides a capability for identification of an eluting compound. On the negative side, the high pressure and flow rate of the SFC and the vacuum requirements of less than 10^{-4} torr of the mass spectrometer require a suitable interface.

The idea of connecting a mass spectrometer to a supercritical fluid chromatograph is not a new one. *Wahrhaftig* and *Giddings* [3] in 1970 and later *Randall* and *Wahrhaftig* [4] in 1978 experimented with a packed column SFC connected to a magnetic sector mass spectrometer. Because of the large volume of supercritical fluid used in the chromatograph, an elaborate nozzle-skimmer system was built to reduce the mobile phase volume and enrich the analyte [4]. **Figure 1** summarizes the *Randall* system. The multi-chamber nozzle-skimmer system with appropriate pumps was able to reduce the SFC column pressure of 65–200 atm and enrich the analyte in the ion source region to a level that mass spectra were obtained for several compounds using nitrogen and ethane, ethylene and carbon dioxide as mobile phases. Electron ionization-like spectra were reported for 1,3,5-triphenylbenzene and anthracene. *Randall* and *Wahrhaftig* also reported the observation of some CO_2 clustering with the analytes and generally poor sensitivity for most compounds. This instrument, although not an analytical prototype, demonstrated that SFC could be successfully coupled to a mass spectrometer and that mass spectra could be obtained.

The advancement of both chromatographic instrumentation and column technology [5, 6] provided the impetus for the present interest in SFC/MS. Capillary

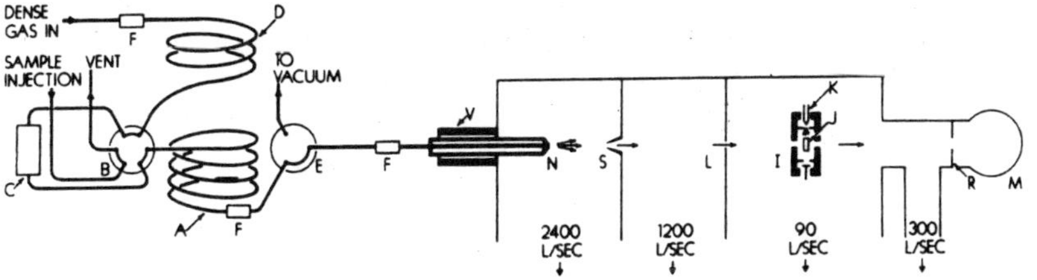

Figure 1
Dense gas chromatograph/mass spectrometer interface. F, filters, sintered stainless steel; D, preheat column; B, sampling valve; C, sample chamber; A, chromatograph column; E, shut-off valve, zero-dead-volume; V, sliding vacuum seal; N, nozzle; S, skimmer; L, collimator; I, ionization chamber; J, ion exit slit; K, filament; R, variable limiting orifice for ionization gauge; M, ionization gauge beam detector. Reprinted by permission from reference [4] (Anal. Chem. **50** (1978) 1708–1705) Copyright 1978, American Chemical Society.

SFC (columns < 75 μm) allowed for the use of microliter flow rates instead of the milliliter flow rates of earlier packed column SFC. The capillary flow rates, when calculated on a STP gas volume, are close to those being used in GC/MS. The pumping capacity of most commercial chemical ionization (CI) mass spectrometers will accept the entire flow from the SFC column without the need of sophisticated skimmer-nozzle systems such as that used by *Randall. Smith* and *Lee* [7] in 1982 described the first direct fluid injection (DFI) interface coupling SFC with a CI mass spectrometer. Since this report, several groups have published on other capillary SFC/MS interfaces and applications along with a simplified packed column interface. A chronological summary of the significant developments of SFC/MS is given in **Table 1**.

Table 1
Chronological Development of SFC/MS.

Group	Accomplishment	Date
Randall	First SFC/MS interface	1978
Smith/Lee	Capillary SFC/MS CI interface	1982
Voorhees et al.	Capillary SFC/MS EI interface	1985
Berry/Games	Moving belt interface	1986
Matsumoto	Vacuum nebulizing interface	1986
Sin/Pang	Pulsed molecular beam	1986
Smith	Packed column interface	1986
Henion	SFC with FTMS	1987

The above reports present the state of the art for SFC/MS ionization modes, interfaces, and forms of MS analyzers. The remainder of this paper will discuss and critically analyze each of these contributions and summarize the authors' assessment of the future use of SFC/MS as a routine analytical procedure.

4.2 Direct Fluid Injection Chemical Ionization Mass Spectrometry

The first practical interface (**Figure 2**) for SFC/MS was introduced in 1982 by *Smith et al.* [7]. The restrictor and probe tip were heated to compensate for decompressive cooling of the fluid. Oven air also was circulated through the DFI probe to assist in maintaining constant temperature conditions. The chemical ionization source was designed to allow axial injection of the SFC effluent (Figure 2), although

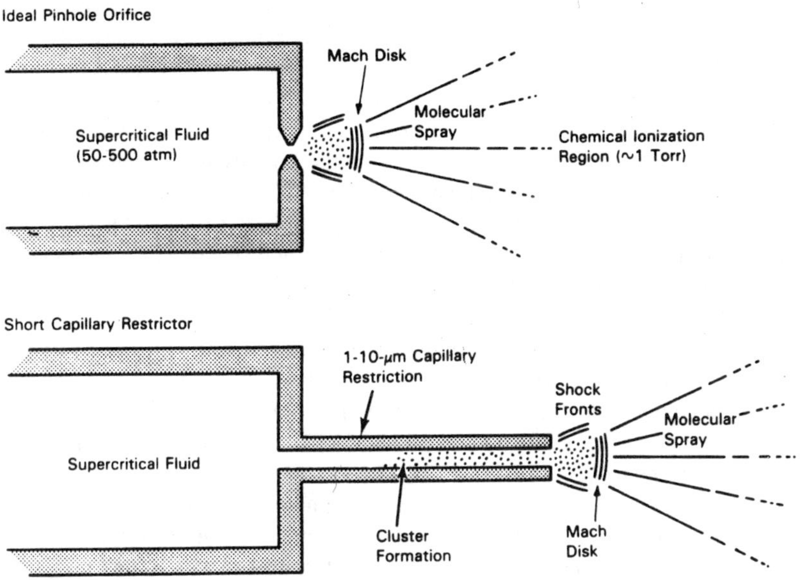

Figure 2
Schematic of the Direct Fluid Injection (DFI) process; pinhole orifice (top) and expansion through a capillary restriction (bottom). Reprinted by permission from reference [7] (Anal. Chem. **54** (1982) 1883–1885) Copyright 1982, American Chemical Society.

tangential injection was also possible. Isobutane and *n*-pentane were used as SFC mobile phases and CI reagents in this early work.

The disadvantages of clustering and poor sensitivity experienced for the molecular beam approach were eliminated by directly introducing the column effluent from capillary columns into the source of the mass spectrometer. Typically gas flow rates for 50 μm i. d. columns range from 0.5–2 ml/min, allowing adequate pumping by most two-staged mass spectrometers. Chemical ionization was used and has been used in the majority of subsequent SFC/MS applications. Simple restrictor devices, such as crimped tubing or laser drilled orifices, have been used to maintain supercritical pressures in the column up to the region immediately adjacent to the ion source. Rapid decompression and expansion through the orifice develops shock fronts and collisional processes in the expanding jet causing breakup of the fluid clusters. Typical CI ion source pressures are in the range of 0.2–1.0 torr. Source pressures in this range provide adequate heat transfer to facilitate the transport of analytes in the vapor phase.

Smith and co-workers have published several examples demonstrating a wide variety of CI-SFC/MS applications. Trichothecene mycotoxins were analyzed [8] using SFC/MS with carbon dioxide mobile phase at 70 and 100 °C. These compounds have been difficult to analyze by either GC or HPLC. GC analysis usually requires derivatization and HPLC detectors generally lack sensitivity. Another example of the benefits of SFC over GC is demonstrated in the analysis of thermally labile carbamate pesticides [9]. The total ion chromatogram for a seven component mixture of acid and carbamate pesticides not attainable by GC is illustrated in **Figure 3**. Carbon dioxide at 75 °C was used in this study as the mobile phase with ammonia and methane as CI reagent gases. While only limited structural information is possible with chemical ionization, different chemical ionization reagent gases have been used to initiate various characteristic fragmentation patterns [8]. Soft ionization with ammonia CI yields a greater selective detection of polar components such as N- or S-substituted compounds due to its higher proton affinity. Alkanes are not detected using ammonia reagent gas making it possible to detect N-containing pesticides in a hydrocarbon matrix.

Smith et al. [10] have reported capillary column SFC separations using rapid pressure programming and short columns ($<$ 2 meters). **Figure 4** compares the results of a rapid separation of a complex coal tar mixture with a high resolution separation. A 15 m capillary column was used for the high-resolution separation and a 1.75 m capillary column was used for the rapid separation. Pressure programming rates approaching 50 bar/min were possible with the shorter columns. The faster program rate compresses chromatographic peaks and increases sensitivity. Rapid separations were shown to be feasible when a decrease in resolution can be tolerated.

Other applications which have been accomplished with SFC/MS in the CI mode are: analysis of polybutadiene polymers [9], PAH in marine diesel fuel [11], and components in bee's wax [12]. Observed sensitivities are generally in the mid-picogram range for full scanning with as low as 0.1 picogram detection limits reported for a series of biphenyl compounds using single ion monitoring [10].

4.3 Other Types of Interfaces Used with Chemical Ionization

Recently, *Smith et al.* [13] reported the development of a high flow rate SFC/MS interface **(Figure 5)**. The design, utilized a mechanical pump behind the CI repeller electrode, which provided improved ionization efficiency and sensitivity when compared to direct pumping of the ion source. However, problems of backstreaming and source contamination are of concern, especially if the flow of gas is interrupted. Ad-

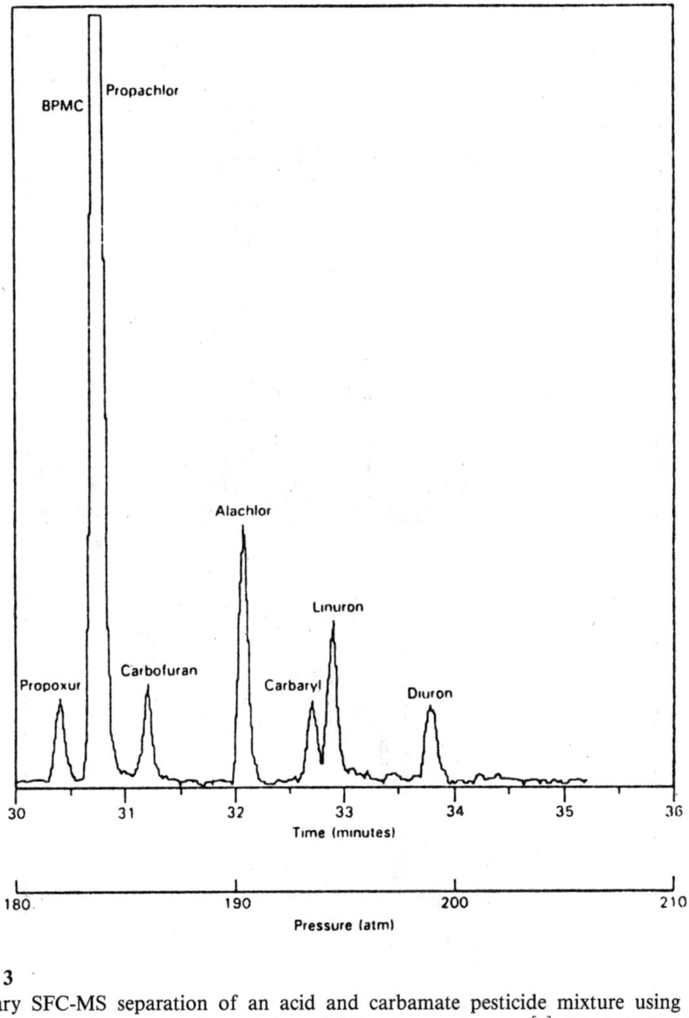

Figure 3
Capillary SFC-MS separation of an acid and carbamate pesticide mixture using CO_2 at 75 °C. Reprinted by permission of the publisher from reference [9] (HRC & CC **9** (1986) 145–153).

ditional pumping allowed splitless injection which improved quantitation by eliminating the inconsistencies which occur with split injection. Microbore and capillary columns were used to separate the nonionic surfactant Triton X-100 polymer. CI reagent gases were introduced directly into the source volume. Poorer

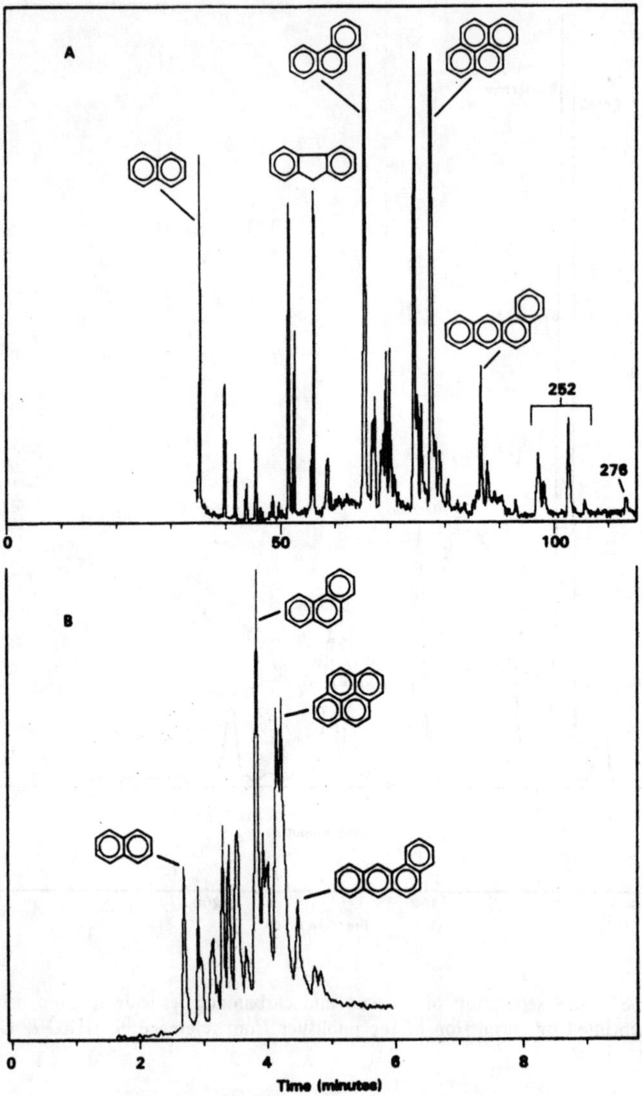

Figure 4
Comparison of a high-resolution capillary SFC/MS separation of a coal tar extract obtained using (A) a 15 m x 50 μm column with (B) a rapid separation on a 1.75 m x 50 μm column. Reprinted by permission from reference [10] (Anal. Chem. **56** (1984) 2476–2480) Copyright 1984, American Chemical Society.

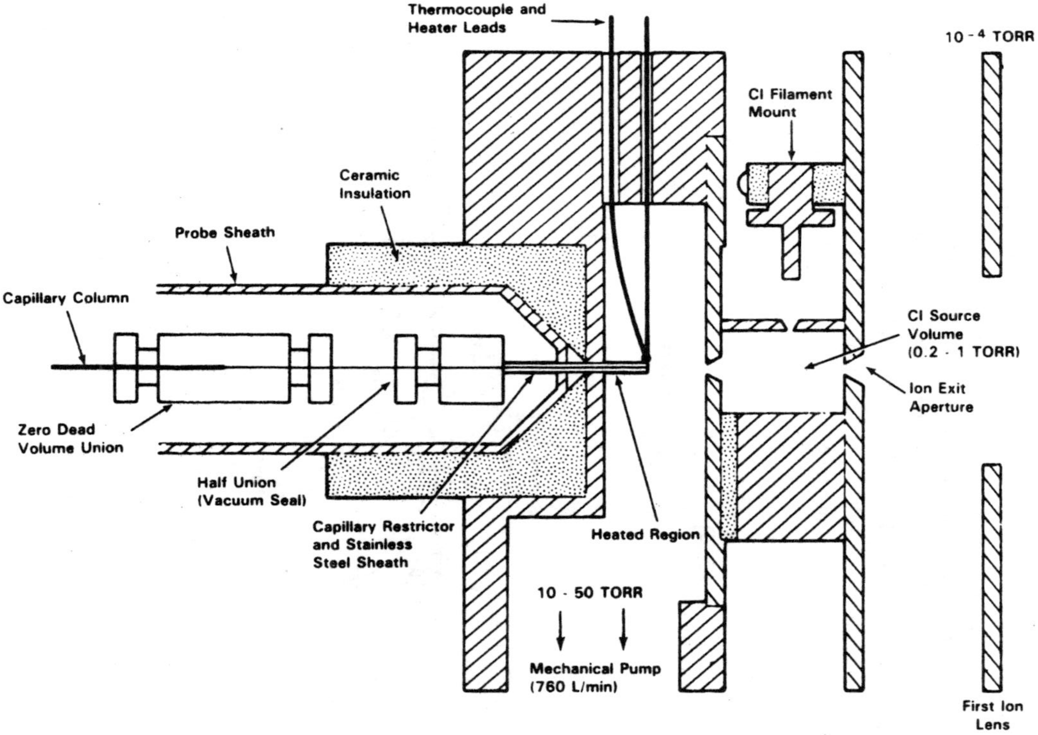

Figure 5
Schematic of the high flow rate (HFR) SFC-MS interface. Reprinted by permission from reference [13] (Anal. Chem. **59** (1987) 13 – 22) Copyright 1987, American Chemical Society.

resolution was observed using microbore columns than that generally obtained with capillary columns.

Another interesting approach for introducing supercritical fluids into the MS source utilizes a pulsed nozzle [14]. Although *Sin et al.* used seeded fluids rather than the effluent from a chromatographic column, their design demonstrates potential SFC/MS applications. Supercritical CO_2 was expanded into a supersonic beam **(Figure 6)** similar to the approach of *Randall* and *Wahrhaftig* [4]. The pulsed nozzle permitted the use of a larger diameter orifice (150 μm) than those used by *Randall* which increased the on-axis density in the jet, and the sensitivity by approximately two orders of magnitude. A tunable UV laser source and time-of-flight mass spectrometer were used with the interface (Figure 6) to obtain parent ion mass spectra

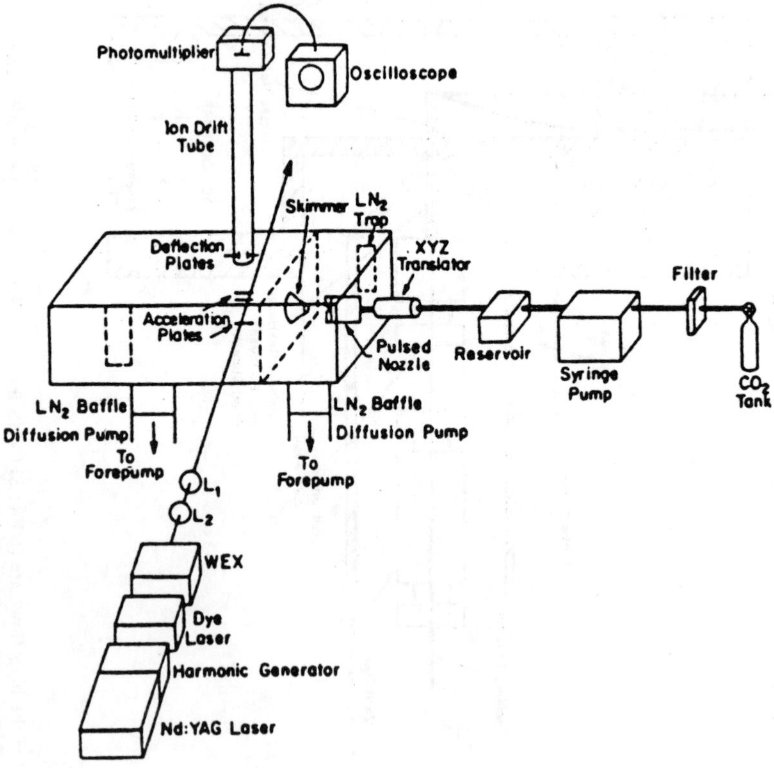

Figure 6
Schematic of Sin pulsed nozzle interface. Reprinted by permission from reference [4] (Anal. Chem. **58** (1986) 487–490) Copyright 1986, American Chemical Society.

of PAH. Extensive cryobaffling was needed and cluster formation was not significant.

Coupling SFC/MS with a self-spouting and nebulizing interface (**Figure 7**) has been reported by *Matsumoto* and *Tsuge* [15]. This technique was originally developed for HPLC/MS and was adapted to SFC with a packed fused silica column interfaced to a quadrupole MS. Helium nebulizing gas was introduced coaxially after UV detection. Desolvation of the nebulized effluent was accomplished by means of an electrical heater operated at about 700 °C. About half of the effluent was diverted through the split vent. Chemical ionization spectra of fat and water soluble vitamins, styrene oligomers, and triglycerides were obtained. Injections of 1.0 μl of the various analytes at concentrations of 1 mg/ml were used, indicating the apparent lack of sensitivity of the instrument. However, the system enables fairly long term stable operation when compared with other designs, and avoids plugging of small diameter orifices which causes erratic flow rates.

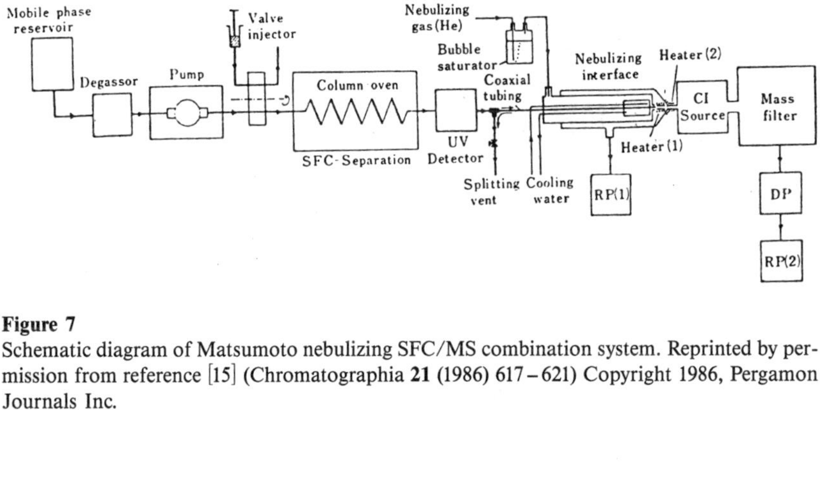

Figure 7
Schematic diagram of Matsumoto nebulizing SFC/MS combination system. Reprinted by permission from reference [15] (Chromatographia **21** (1986) 617–621) Copyright 1986, Pergamon Journals Inc.

4.4 Direct Fluid Injection Electron Ionization Mass Spectrometry

Smith [16] first reported obtaining EI spectra using a DFI interface. The reported sensitivity for EI was less than previously reported for CI analysis, however, the problems of obtaining EI spectra were not defined. Two papers [17, 18] have been published on EI-SFC/MS by *Voorhees et al.* which present a broader perspective of the problems and limitations of EI SFC/MS. On using a DFI interface, similar to those described by *Smith*, significant problems occurred with clustering of the CO_2

with a combined EI/CI source. The cluster intensities could be minimized (**Figure 8**) by optimizing the source temperature and pressure (via flow rate). Although the sensitivity in using this arrangement was not satisfactory (the limit of detection was of the order of 100 ng for triglycerides), EI spectra of large intact lipids were gathered (**Figure 9**) including that of a diphytanyl glycerol diether (MW = 652). The reported spectrum represented the first EI spectrum of that intact molecule [17].

The poor sensitivity was attributed to the high pressure of the CO_2 in the source region. To minimize the pressure, an open source was constructed. In addition, with the development of the "integral restrictor" by *Guthrie* [19], a simplified capillary GC-like interface could be used. The design of the source and interface is shown

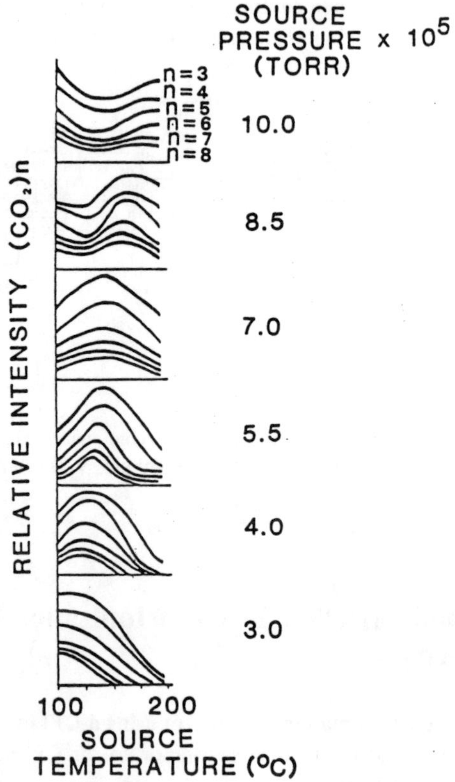

Figure 8
CO_2 cluster intensities as a function of ion source pressure and temperature. Reprinted by permission of the publisher from reference [17] (HRC & CC **8** (1985) 528–531).

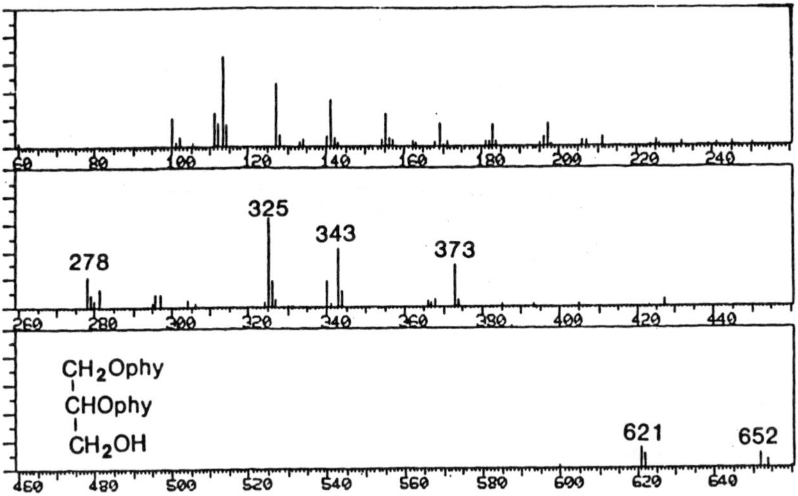

Figure 9
EI mass spectrum of diphytanylglycerol diether. Reprinted by permission of the publisher from reference [17] (HRC & CC **8** (1985) (528–531).

in **Figure 10**. A quartz tube (3 cm x 0.5 mm i. d.), drawn down to a jet at one end was appended to the glass GC/MS interface (Allen Scientific, Boulder, CO). While the GC/MS interface was maintained at the same temperature as the SFC column, the quartz jet was independently heated with a wrapped wire heater. The column was inserted through the GC/MS interface and quartz heater such that the column end (restricted) was approximately 1 cm from the jet tip. This allowed the column effluent to be heated for improved decomposition of the solvent clusters prior to entering the ionization region. The open source configuration and injection perpendicular to the quadrupole axis allowed more efficient pumping of the CO_2 effluent.

The quartz jet was actually heated differentially such that a temperature gradient was established. Typically the jet operated at 300 °C at the base (nearest the GC interface) and 400–450 °C at the jet tip. This allowed for on-column heating of the mobile phase just prior to decompression where heat transfer is more efficient than at the point of decompression. The mass spectrometer source temperature was set at 250 °C. Increasing the source temperature did not increase the sensitivity for the compounds of interest.

The sensitivity of this interface for EI-SFC/MS is demonstrated in the example in **Figure 11**. The separation of a mixture with 30 ng per component, except for

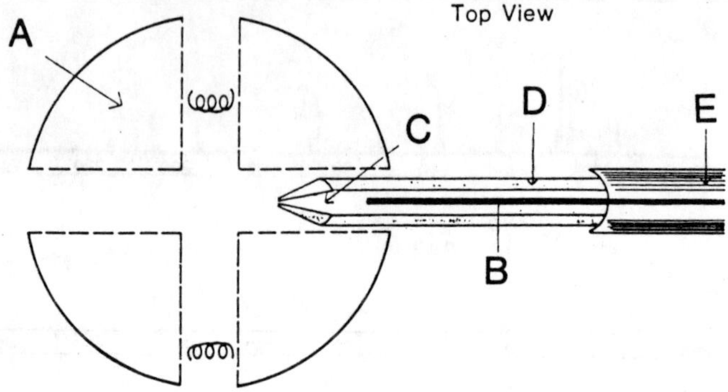

Figure 10
Schematic diagram of EI-SFC/MS interface: (A) EI source; (B) capillary column; (C) quartz jet heater; (D) insulating quartz sheath; (E) GC/MS interface. Reprinted by permission of the publisher from reference [18] (HRC & CC **10** (1987) 100–101).

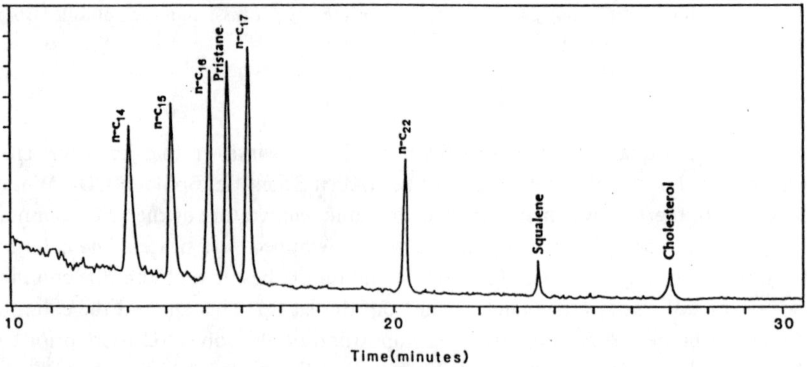

Figure 11
Total ion chromatogram for SFC/MS of an eight component mixture. CO_2 linear pressure program: 100–300 atm in 30 min; oven temperature: 120 °C. Reprinted by permission of the publisher from reference [18] (HRC & CC **10** (1987) 100–101).

cholesterol (50 ng) was accomplished in less than 30 minutes on the 6 m column. Peak tailing occurred for some of the hydrocarbons which eluted on the solvent tail, and poor peak shape indicated that column overloading may have occurred for cholesterol. Detection limits of 1–5 ng were observed for total ion monitoring, while detection limits below 100 pg were easily achieved using single ion monitoring.

When the same mixture was analyzed by a GC/MS using a capillary column coupled to the MS via the same interface (using He as carrier gas), the detection limit was markedly improved (**Figure 12**). The decreased sensitivity under SFC conditions was attributed to the high detector background noise when compared to GC. The pressure in the analyzer-detector region was about one order of magnitude greater for SFC than GC, thus causing a significant increase in detector noise. Some loss of sensitivity could also be attributed to decreased ionization efficiency with increased SFC source pressures. Even with the more open source design, source pressures were greater for SFC than GC because of reduced pumping efficiency for CO_2. Higher source pressure can also decrease the penetration of electrons and affect the ionization efficiency. Therefore, pressure programming CO_2 in an SFC separation could result in decreased ionization efficiency of later eluting compounds. This may actually be the case, as indicated by the lower relative response to squalene (Figure 11) and cholesterol in the SFC chromatogram, compared to GC separation (Figure 12). A small improvement in the sensitivity was achieved by using 110 eV electron energy rather than 70 eV.

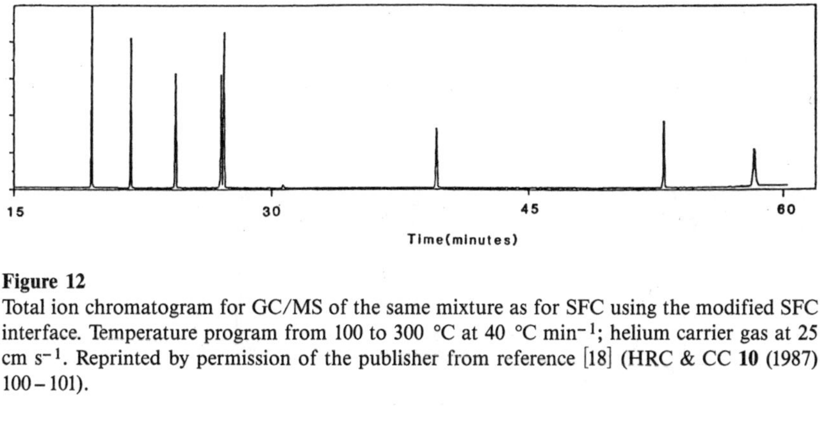

Figure 12
Total ion chromatogram for GC/MS of the same mixture as for SFC using the modified SFC interface. Temperature program from 100 to 300 °C at 40 °C min^{-1}; helium carrier gas at 25 cm s^{-1}. Reprinted by permission of the publisher from reference [18] (HRC & CC **10** (1987) 100–101).

The EI spectrum of cholesterol (**Figure 13**) demonstrates the absence of CO_2 clusters. The highest mass peak of cholesterol (m/z = 368) corresponds to the $(M-H_2O)+$ ion. The molecular ion was not observed at the operating temperature of the quartz jet (300–400 °C). By lowering the temperature of the quartz jet to 200–250 °C, the molecular ion was detected but the sensitivity and quality of the spectrum were greatly reduced. Solvent clusters also became apparent below 250 °C. **Figure 14** shows a total ion chromatogram of a seven component test mixture having 10 ng per component. Again, limits of detection are in the 1–5 ng range.

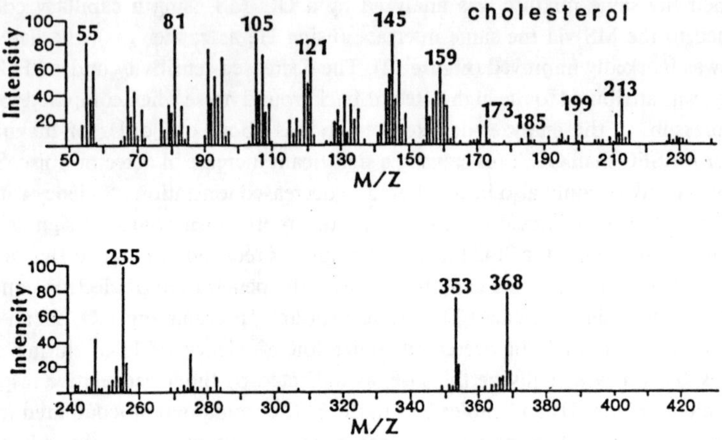

Figure 13
SFC/MS spectrum of cholesterol. Reprinted by permission of the publisher from reference [18] (HRC & CC **10** (1987) 100–101).

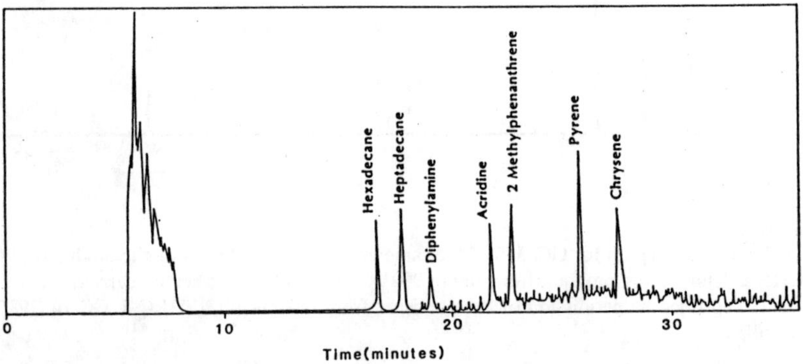

Figure 14
Total ion chromatogram for SFC/MS of a seven component mixture having 10 ng per component. CO_2 linear pressure program: 100–300 atm in 30 min; oven temperature: 120 °C. Reprinted by permission of the publisher from reference [18] (HRC & CC **10** (1987) 100–101).

Recently, the advantages of interfacing SFC with Fourier-transform mass spectrometry (FTMS) using electron ionization have been documented by *Henion et al.* [20]. Because of its ion trapping capabilitities, acceptable EI detection limits, as well as accurate mass measurements, are possible using this approach. These advantages are not attainable using direct liquid introduction HPLC/FTMS. Chemical ionization as well as EI spectra are possible from this instrument.

A capillary SFC column was interfaced to the FTMS into a differentially pumped dual cell. The two cells share a common ion trapping plate and the superconducting magnet was operated at a field strength of 3 tesla. A modified GC/MS transfer line was maintained at a nominal temperature of 100 °C using a resistive heater.

Continuous introduction of supercritical CO_2 was demonstrated even though operating pressures were considered high for FTMS. Nevertheless, accurate mass measurements and EI mass spectra were obtained with 25 ng of a caffeine standard (the TIC is shown in **Figure 15**).

The lack of sufficient temperature control of the transfer line region and selectively heating the restrictor were major limitations. Improvements in the interface design may extend the range of applications to include less volatile analytes. The availability and complexity of FTMS instrumentation may delay development of practical SFC/FTMS applications. Analysis of complex mixtures remains to be demonstrated.

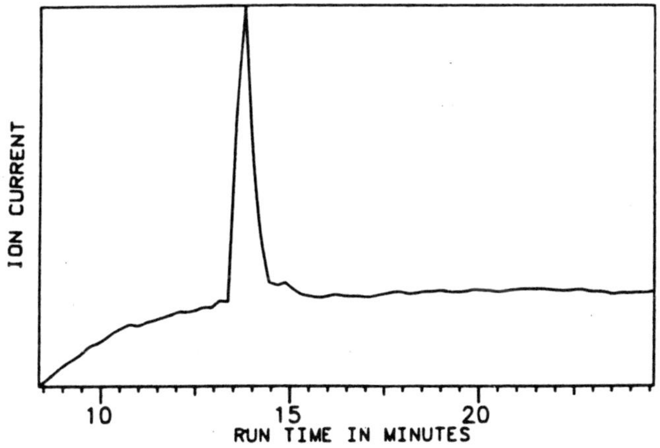

Figure 15
(A) Total ion current profile for 25 ng of caffeine on column. Reprinted by permission from reference [20] (Anal. Chem. **59** (1987) 1309–1312) Copyright 1987, American Chemical Society.

4.5 Commercial Interfaces

Commercial HPLC instrumentation may also have applications for SFC/MS. *Berry* and *Games* [21] have used a moving belt HPLC/MS interface with a thermospray deposition device to separate xanthines, sulfonamides, and alkaloids. **Figure 16** shows the TIC trace of a complex ergot alkaloid extract. Several of the alkaloids were identified by EI-SFC/MS spectra. Packed columns were used in this study so effluent splitting of 80–90% was necessary. Peak broadening also occurred due to the dead volume in the connection between the SFC and the moving belt. Whereas the SFC/MS analysis of the reported mixture was accomplished in 10 min, HPLC/MS required over 1 h. The improved chromatographic characteristics of SFC allowed for separation of some components which were not identified by HPLC. This approach enabled both EI and CI spectra to be obtained.

A commercial HPLC/MS interface (Therma Beam ™) available from Extrel has been evaluated for SFC/MS application and preliminary results were encouraging [21]. The interface thermally nebulizes the column effluent generating submicron particles which pass through two stages of momentum separation before entering the ion source. Because of the proprietary nature of the momentum separator particle beam interface, details of its design are not included. However, an example of

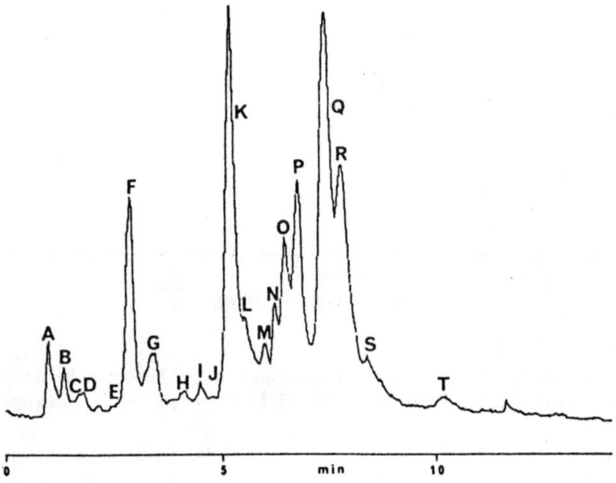

Figure 16
TIC of an ergot alkaloid extract obtained by EI-SFC/MS with a moving belt interface. Reprinted by permission of Elsevier Scientific Publishers from reference [21] (J. Chromatogr. **363** (1986) 147–158).

a TIC using EI-SFC/MS for a fish oil extract is demonstrated in **Figure 17**. Reasonable separation of the triglycerides was possible using a packed column. The interface offers good EI performance characteristics with minimal clogging difficulties. EI spectra for eleven triglycerides enabled compounds to be identified with molecular weights in excess of 1000 daltons. The interface can also be used in the CI mode of operation using a variety of reagent gases.

4.6 Summary

The DFI interface is the simplest and offers acceptable sensitivity in the CI mode. This type of interface employed with EI provides much less sensitivity, which in part might be improved with a better ion source design. Small diameter restrictors are used with the DFI interface to maintain supercritical conditions up to the ion source. The restrictors are a potential source of plugging and erratic performance.

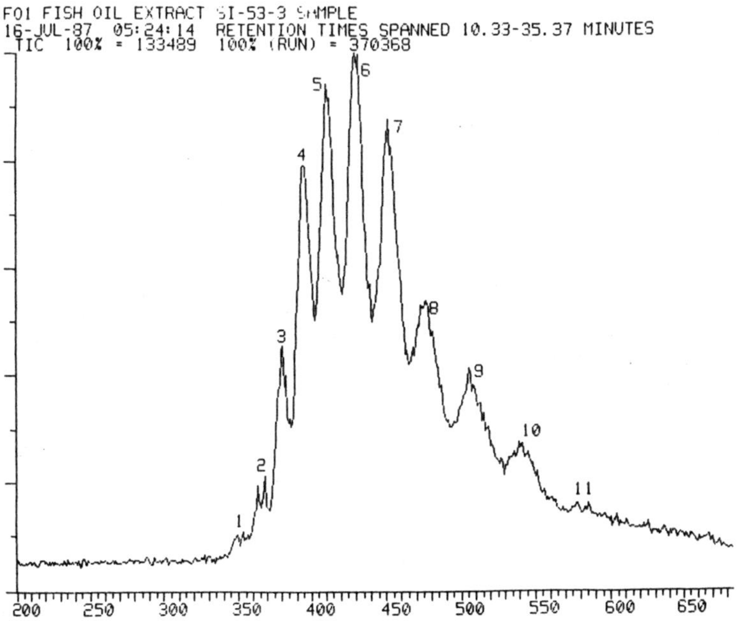

Figure 17
TIC using EI-SFC/MS for a Fish Oil Extract. Reprinted by permission of Extrel from reference [22]. All rights reserved by Extrel.

Henion [20] has used the ion collection capabilities of Fourier Transform MS to overcome some of the EI sensitivity problems. This type of an approach along with other possibilities such as charge transfer ionization might allow EI-type spectra to be obtained with the same sensitivity as CI.

There are several existing technologies which should be applied to SFC/MS research. Atmospheric pressure (API) mass spectrometry offers the possibility of using larger restrictors and higher gas flow rates. This would eliminate the problems of clogging and also allow the use of packed columns. The only major drawback to API is the limitation of not being able to use electron ionization. However, the use of tandem MS might be a route to obtain fragmented spectra which can be used for identification. The area of tandem MS should also receive additional attention with existing CI work.

Supercritical fluid chromatography/mass spectrometry at the present time is not a routine technique available to most analytical laboratories. The area of SFC/MS is at about the same development stage as SFC when *Lee* published his first paper on capillary SFC. The advancement of SFC has occurred because of the commercial interests. For SFC/MS to reach its full potential, there will have to be commercial equipment available. With the entry of two or three manufacturers, the future of SFC/MS appears bright.

References

[1] *M. Novotny, S. R. Springston, P. A. Peaden, G. C. Fjeldsted,* and *M. L. Lee,* Anal. Chem. **53** (1981) 407A–414A.

[2] *R. D. Smith, B. W. Wright,* and *H. R. Udseth,* "Capillary Supercritical Fluid Chromatography and Supercritical Fluid Chromatography-Mass Spectrometry" in Chromatography Separation Chemistry, ACS Symposium Series 297 (1986) p. 260–293.

[3] *J. C. Giddings, M. Myers,* and *A. L. Wahrhaftig,* Int. J. Mass Spectrom. Ion Phys. **4** (1970) 9–20.

[4] *L. G. Randall* and *A. L. Wahrhaftig,* Anal. Chem. **50** (1978) 1703–1705.

[5] *L. Blomberg* and *T. Wännman,* J. Chromatogr. **168** (1979) 81–88.

[6] *P. A. Peaden, J. C. Fjeldsted, M. L. Lee, S. R. Springston,* and *M. Novotny,* Anal. Chem. **54** (1981) 1090–1093.

[7] *R. D. Smith, W. D. Felix, J. C. Fjeldsted,* and *M. L. Lee,* Anal. Chem. **54** (1982) 1883–1885.

[8] *R. D. Smith, H. R. Udseth,* and *B. W. Wright,* J. Chromatogr. Sci. **23** (1985) 192–199.

[9] *B. W. Wright, H. T. Kalinoski, H. R. Udseth,* and *R. D. Smith,* HRC & CC **9** (1986) 145–153.

[10] *R. D. Smith, H. T. Kalinoski, H. R. Udseth,* and *B. W. Wright,* Anal. Chem. **56** (1984) 2476–2480.

[11] *B. W. Wright, H. R. Udseth, R. D. Smith,* and *R. N. Hazlett,* J. Chromatogr. **314** (1984) 253–262.

[12] *S. Hawthorne* and *D. J. Miller,* J. Chromatogr. **388** (1987) 397–409.

[13] *R. D. Smith* and *H. R. Udseth,* Anal. Chem. **59** (1987) 13 – 22.
[14] *C. H. Sin, H. M. Pang, D. M. Lubman,* and *J. Zorn,* Anal. Chem. **58** (1986) 487 – 490.
[15] *K. Matsumoto* and *S. Tsuge,* Chromatographia **21** (1986) 617 – 621.
[16] *R. D. Smith, H. R. Udseth,* and *H. T. Kalinoski,* Anal. Chem. **56** (1984) 2971 – 2973.
[17] *G. Holzer, S. DeLuca,* and *K. J. Voorhees,* HRC & CC **8** (1985) 528 – 531.
[18] *S. D. Zaugg, S. J. DeLuca, G. U. Holzer,* and *K. J. Voorhees,* HRC & CC **10** (1987) 100 – 101.
[19] *E. J. Guthrie* and *H. E. Schwartz,* J. Chromatogr. Sci. **24** (1986) 236 – 241.
[20] *E. D. Lee, J. D. Henion, R. B. Cody,* and *J. A. Kinsinger,* Anal. Chem. **59** (1987) 1309 – 1312.
[21] *A. J. Berry, D. E. Games,* and *J. R. Perkins,* J. Chromatogr. **363** (1986) 147 – 158.
[22] *J. D. Buchner* and *R. C. Willoughby,* "Interfacing Techniques for SFC/MS," Paper # 265, Rocky Mountain Analytical Symposium, Denver, CO, 1987.

5 Capillary SFC and SFC/FT-IR with Polar Mobile Phases

Gilbert L. Pariente and Peter R. Griffiths
Department of Chemistry, University of California, Riverside, CA 92521–0403, USA

Gilbert L. Pariente is a doctoral candidate in the Department of Chemistry of the University of California, Riverside. He received his B. A. degree from the University of Virginia in 1982 and will shortly obtain his Ph. D. degree. Besides his interests in SFC, he has also worked in the areas of supercritical fluid extraction and vibrational spectrometry. He has designed and constructed a computer-controlled pneumatic amplifier pump and is currently building a stepped-scanning interferometer for photoacoustic spectrometry. He will start work with the Sherwin-Williams Company, Chicago, in March, 1988.

Peter Griffiths received his B. A. in chemistry from Oxford University in 1964 and his D. Phil. from the same institution three years later. After two years postdoctoral research at the University of Maryland with the late *Ellis Lippincott*, he became product specialist for FT-IR spectrometry at Digilab, Inc., in Cambridge, MA. One year later, he was appointed manager of analytical services at Sadtler Research Laboratories in Philadelphia, PA. In 1972, he joined the faculty of Ohio University, becoming full professor of analytical chemistry in 1978. In September 1982, he joined the chemistry department at the University of California, Riverside.

Professor Griffiths' current research interests include the development of the GC/FT-IR, HPLC/FT-IR, and SFC/FT-IR interfaces, application of diffuse reflectance infrared spectrometry to the chemistry of coals and surfaces, development of optical subtraction and polarization modulation techniques for analytical chemical problems, refinement of data processing techniques through the use of Fourier transforms, the development of remote sensing techniques by infrared emission spectrometry, and the use of supercritical fluids for selective extractions from complex materials.

In 1977 *Professor Griffiths* received the Coblentz Award for molecular spectroscopy and he was presented the 1985 Spectroscopy Society of Pittsburgh Award.

Summary:

The use of polar mobile phases in capillary SFC allows the elution of more polar analytes than CO_2 at the same density. Contrary to earlier prognostications, the Hildebrand solubility parameter does not allow the relative eluent strength of polar supercritical fluids to be predicted.

Conventional SFC/FT-IR techniques in which the column effluent is passed through a high-pressure flow-cell are inappropriate for polar fluids in the supercritical state as absorption of the incident radiation is too strong over much of the useful infrared spectrum. Elimination of the mobile phase prior to the measurement of the spectrum permits this limitation to be overcome. Interpretable spectra can be obtained from eluites injected in a quantity of a few nanograms over the complete infrared spectral range using this technique.

Concentrate:

Capillary supercritical fluid chromatographic separations of polar molecules using CCl_2F_2 as the mobile phase are described. CCl_2F_2 appears to be a stronger solvent than CO_2 at the same density. Infrared spectra of eluites can be obtained at the low nanogram level after rapid elimination of the mobile phase.

5.1 Introduction

It is often noted that supercritical fluid chromatography (SFC) should exhibit the combined advantages of gas chromatography (GC) and high performance liquid chromatography (HPLC). However, the great progress in capillary supercritical fluid chromatography (SFC) which has been made since the description of this technique by *Novotny et al.* [1] in 1981 has largely resulted from advances in column technology rather than through the use of novel mobile phases. Better methods of crosslinking siloxane-based liquid phases have resulted in improved stability on increasing the temperature of the column and the density of the mobile phase [2]. Siloxanes with a wide variety of substituents, including methyl, *n*-octyl, octadecyl, phenyl, biphenyl, and cyanoalkyl groups, have been prepared. Recently, liquid crystalline stationary phases have been successfully used for the separation of chiral molecules. A careful choice of siloxane stationary phase with the correct substituent allows the selectivity of many separations to be controlled in both GC and capillary SFC.

The wide variety of stationary phases described above is largely a result of developments in the field of GC, as the only fundamental difference between capillary GC and SFC columns (aside from the necessity of reducing the internal diameter) is the need to immobilize the stationary phase for SFC by improved crosslinking. In gas chromatography the mobile phase has little or no effect on the capacity factor for the elution of a given analyte, k', or the selectivity coefficient for the separation of two similar molecules, α. Generally the only reasons to choose a carrier gas other than helium are the need to perform faster analyses (*e. g.* with H_2) or the need to use N_2 for an electron capture detector. For HPLC, on the other hand, the role of the mobile phase is as important in determining k' and α as that of the stationary phase. Because solutes in SFC *dissolve* in the mobile phase, it is logical to believe that the polarity of the mobile phase in SFC should play at least as large a role in determining retention behavior as the polarity of the stationary phase.

This is certainly the case for SFC separations effected using packed columns, where polar modifiers such as methanol, propylene carbonate, or formic acid are commonly added to CO_2 at concentrations between 0.1 and 10%. The main function of the modifier in packed-column SFC is to block active sites on the surface of the stationary phase. This function is less relevant to capillary SFC where the fused silica surface is both less available to the solutes and less active than the surface of silica particles in packed columns. The second function of the polar modifier is to increase the solubility of polar molecules in the mobile phase in order to reduce the enthalpy of transfer of the solutes between the stationary and mobile phase, and hence to facilitate the elution of polar molecules. This function should be equally applicable to capillary SFC. In this paper we examine the effect of using several polar fluids as mobile phases for capillary SFC, with particular emphasis on the use of dichlorodifluoromethane.

Giddings [3] has noted a correlation between the elution power, ϵ°, of conventional liquid solvents in liquid chromatography with alumina packings and the Hildebrand solubility parameter, δ, and he suggested that δ might be a good measure of solvent strengths of dense gases (supercritical fluids). Under this criterion, CCl_2F_2 (for which $\delta \approx 7.9$) would be expected to be a much poorer solvent than either CO_2 ($\delta \approx 10.7$, about the same as 2-propanol and pyridine) or even argon ($\delta \approx 8.7$). Surprisingly, SO_2, for which $\delta \approx 11.0$, would not be forecast to be a significantly better solvent for SFC than CO_2, despite its large dipole moment. It should be noted, however, that *Hildebrand* and *Scott* originally cautioned that it is dangerous to employ solubility parameters for polar liquids [4], and a careful comparison shows that there are too many anomalous cases for the correlation of ϵ° and δ to be valid, even for liquid solvents. For example, δ for diethyl ether (7.40) is less than that for cyclohexane ($\delta = 8.2$), but the difference in their ϵ° values is 0.34, showing that diethyl ether is about one thousand times *stronger* as a solvent than cyclohexane. Thus we believe that it is highly unlikely that a solubility parameter calculated for polar supercritical fluids is likely to be of any use in forecasting their behavior as mobile phases for capillary SFC. Indeed, the best way of determining the relative strengths of polar and nonpolar fluids for SFC appears to be by direct observation.

The relevant properties of several potential candidates for SFC with a polar mobile phase are summarized in **Table 1**. These molecules were selected on the basis of a non-zero dipole moment and values of the critical temperature, T_c, and critical pressure, P_c, which could be expected to be able to be reached in practice without too much difficulty. (CO_2 and methanol were included for reference.) Since T_c is approximately 1.5 T_b, where T_b is the boiling temperature (Kelvin), the critical temperature of most polar organic molecules (where $T_b > 350$ K) is above 200 °C, excluding them as candidates for SFC.

Of the compounds listed, the dipole moments of CO, NO, and N_2O are each less than 0.2, and little benefit would be expected to be gained from their use over the use of CO_2. Certainly, N_2O, the only one of these compounds to have received significant attention for SFC, does not lead to greatly different retention parameters than CO_2 in capillary SFC [5].

Of the others, ammonia has often been believed to have the most potential in view of its high dipole moment and reasonably low T_c [6]. It is undoubtedly true that the solubility of many polar compounds in liquid ammonia is much greater than their solubility in supercritical CO_2. Thus the use of supercritical NH_3 leads to decreased retention of molecules such as large polycyclic aromatic hydrocarbons in capillary SFC in comparison to retention times found with CO_2. However, super-

Table 1
Relevant properties of polar fluids

	T_c (°C)	P_c (atm)	Dipole moment (Debyes)
Ammonia	132.5	112.5	1.47
Carbon dioxide	31	72.9	0.000
Carbon monoxide	-140	34.5	0.112
Hydrogen chloride	51.4	82.1	1.08
Hydrogen cyanide	183.5	48.9	2.98
Nitric oxide	-93	64	0.153
Nitrous oxide	36.5	71.7	0.167
Sulfur dioxide	157.8	77.7	1.63
CHF_3	25.9	46.9	1.65
CCl_2F_2	111.5	39.6	0.51
CCl_3F_3	28.8	38.2	0.50
Dimethyl ether	127	52.6	1.30
Methanol	240	78.5	1.70

critical NH_3 has been shown to react with siloxane linkages, and most of the siloxane stationary phases used for capillary SFC break down under prolonged exposure to this fluid.

HCN has a rather high T_c, but its dipole moment is almost twice as high as any other compound listed making it an attractive candidate for SFC. However, its severe toxicity will probably preclude routine use of HCN as a mobile phase for SFC. The critical parameters for HCl are not too different from those of CO_2, but it is likely that both the corrosive and toxic properties of HCl could preclude its use for SFC. Sulfur dioxide also has apparently favorable properties, and we have investigated the feasibility of using SO_2 as the mobile phase for capillary SFC.

Of the organic compounds listed, the use of dimethyl ether for capillary SFC has not been reported but appears feasible. Methanol has been used as a polar modifier for CO_2 for both packed column [7] and capillary [8] SFC, but its T_c is much too high for routine SFC using the neat fluid. Although the elution of polar compounds from capillary columns is speeded up on addition of methanol to supercritical CO_2 [8], the effect is not as great as it is for separations made using silica-packed columns. It is also noteworthy that if neat CH_3OH is used above its P_c and T_c for packed capillary SFC using fused silica tubing, methanol not only "dissolves" the silica packing very rapidly, presumably with the formation of $Si(OCH_3)_4$, but also will then attack the fused silica tubing [9]! On the other hand, the chlorofluorocarbons appear to be promising candidates and we have started to investigate their potential as mobile phases for capillary SFC. Some preliminary observations on the use of CCl_2F_2 for capillary SFC will be reported in the following section.

5.2 Instrumental Requirements with CCl_2F_2

During previous work with strongly polar analytes, we observed that the presence of carboxylic acid groups led to greatly increased retention on most columns when supercritical CO_2 was used as the mobile phase [10]. Although monocarboxylic acids could sometimes be eluted with CO_2 provided that the molecule did not also possess many other OH or NH_2 groups, it was quite rare that dicarboxylic acids were not completely retained with CO_2 as the mobile phase, no matter which stationary phase was employed. Even such a simple molecule as isophthalic acid (benzene, 1, 3-dicarboxylic acid) could not be eluted from a DB-5 (5% diphenyl – 95% dimethyl polysiloxane) column. This behavior may be because of the rather low solubility of these polar molecules in supercritical CO_2 since the free energy of solvation is insufficient to overcome the strong intermolecular hydrogen bond network of this dicarboxylic acid and other polycarboxylic acids such as trimesic acid (benzene, 1,3,5-tricarboxylic acid).

For this investigation, the pump was a computer-controlled pneumatic amplifier [11]. Pressure-temperature-density relationships were calculated by the method of *Lee* and *Kesler* [12]. Fused silica capillary columns were mounted in the oven of a Perkin-Elmer Model 3920 gas chromatograph. All fluorocarbons give a large signal on a flame ionization detector (FID). However, these compounds have very little absorption in the ultraviolet at wavelengths longer than 214 nm. We therefore modified a Waters Model 440 ultraviolet absorption detector, which was originally designed as a detector for HPLC, for use with capillary SFC.

It was recognized that the volume of the cell had to be less than about 200 nL in order that peak broadening by the detector was minimized. The column, which was of 100 μm internal diameter (i. d.) and 350 μm outside diameter (o. d.), was fed through a length of 410 μm i. d. fused silica tubing and sealed in place with an epoxy adhesive. A second length of 77 μm i. d., 375 μm o. d. tubing was fed through in the opposite direction and sealed so that there was a gap of about 1 mm between it and the column. The pressure restrictor was located at the other end of this tube. The polyimide coating was then burnt off the fused silica "sleeve", and the tubing was then mounted in the detector in place of the original 1-cm cell, so that the gap between the column and the restrictor tubes was located at the same height as the aperture of the original cell. A 600 μm diameter hole allowed light to pass through the cell. While this arrangement clearly allows some stray light, the sensitivity was high enough to obtain an adequate signal when less then 1 ng of most analytes with a suitable chromophore, was injected. The volume of the tubing being observed was approximately 90 nL. Since the volume of the injection loop was 60 nL, the volume of the detector cell was not expected to degrade the chromatographic resolution by a significant amount.

5.3 Retention Behavior with CCl_2F_2 as the Mobile Phase

As noted above, polycarboxylic acids are often too insoluble in CO_2 to permit elution by capillary SFC. For example, isophthalic acid is not able to be eluted by CO_2 at a density of 0.6 g ml^{-1} from a DB-5 column at 70 °C in a reasonable time ($k' > 20$). In contrast this molecule was able to be eluted from the same type of column at 160 °C by CCl_2F_2 at a density of 0.3 g ml^{-1} with a k' of 3.9. Apparently the solvation of this molecule by CCl_2F_2 was strong enough that the intermolecular hydrogen bonds were able to be broken. A detailed study of the enthalpy of transfer of selected molecules between a DB-5 stationary phase and several mobile phases was initiated, and preliminary results are reported below.

As noted by *Yonker* and *Smith* [13], solute retention in SFC can be defined by the thermodynamic relationship:

$$\Delta G_T^\circ = -RT \ln K_D \tag{1}$$

where ΔG_T° refers to the change in Gibbs free energy of solute transfer from unit activity in the mobile phase to a (hypothetical) unit activity of solution in the stationary phase. K_D is the solute distribution coefficient, R is the gas constant and T is the temperature. The thermodynamic relationship of ΔG_T° and the enthalpy of solute transfer between the two phases at infinite dilution, ΔH_T°, and the change in entropy of solute transfer, ΔS_T°, is:

$$\Delta G_T^\circ = \Delta H_T^\circ - T\Delta S_T^\circ \tag{2}$$

Since $K_D = k'/\phi$, when k' is the capacity factor and ϕ is the volumetric phase ratio, we have:

$$\ln k' = \Delta H_T^\circ/RT + (\Delta S_T^\circ/R + \ln \phi) \tag{3}$$

The slope of a plot of $\ln k'$ against $1/T$ at constant density is therefore $-\Delta H_T^\circ/R$, from which the enthalpy of solute transfer may readily be derived. It should be stressed that these measurements must be made at constant density and not constant pressure. In the latter case there is no reason to predict that $\ln k'$ would vary linearly with $1/T$.

Yonker and *Smith* [13] published plots of ln k' *vs* $1/T$ for several solutes eluting through capillary columns coated with two stationary phases (OV-17 and SE-54) using CO_2 as the mobile phase. They showed that ΔH_T° decreased as fluid densities increased over a range of 0.20 to 0.50 g ml^{-1}, indicating the effect of the increased solute-solvent interactions which occur as fluid density increases. We performed a similar study for several compounds eluted from DB-5 (5% diphenyl – 95% dimetylsiloxane) and Carbowax (polyethylene glycol) columns (J & W Scientific). In both cases we planned to inject a test mixture containing benzoic acid, isophthalic acid, trimesic acid, and their methyl esters. However, in trial runs it was found that trimesic acid was totally retained on both columns, possibly as a result of its insolubility in CCl_2F_2 at the density used for these measurements (0.3 g ml^{-1}). For the strongly polar column, neither trimesic acid nor isophthalic acid was eluted. Therefore the test mixtures injected were corrected accordingly.

The elution order for the DB-5 column was methyl benzoate, benzoic acid, dimethyl isophthalate, trimethyl trimesate, isophthalic acid. This appears to be the order of boiling points of these compounds, although not all boiling points are measurable at atmospheric pressure. The elution order is the same as the order of melting points which usually follows the order of boiling points. It is not the order that would be expected from a typically reversed-phase HPLC separation. Conversely, the order elution from the Carbowax column is the order that would be predicted for a normal-phase HPLC separation, *i. e.* methyl benzoate, dimethyl isophthalate, trimethyl trimesate, benzoic acid.

Plots of ln k' *vs* $1/T$ for the two columns are shown in **Figures 1** and **2** and the ΔH_T° values determined from the slopes of these plots are summarized in **Table 2**. The large difference in the enthalpy of solute transfer of benzoic acid for DB-5 and Carbowax indicates a strong hydrogen-bonding interaction between carboxylic acid groups and the polyethyleneglycol stationary phase. The strong retention of isophthalic acid on this column in comparison with that for the DB-5 column is therefore not unexpected.

Table 2
Enthalpies of solute transfer for polar molecules between CCl_2F_2 at 0.3 g ml^{-1} and DB-5 and Carbowax

Probe molecule	$-\Delta H_T^\circ$ DB-5	(kcal $mole^{-1}$) Carbowax
Benzoic acid, methyl ester	5.90	6.90
Isophthalic acid, dimethyl ester	6.99	7.63
Trimesic acid, trimethyl ester	9.70	8.13
Benzoic acid	5.97	9.59
Isophthalic acid	8.97	–

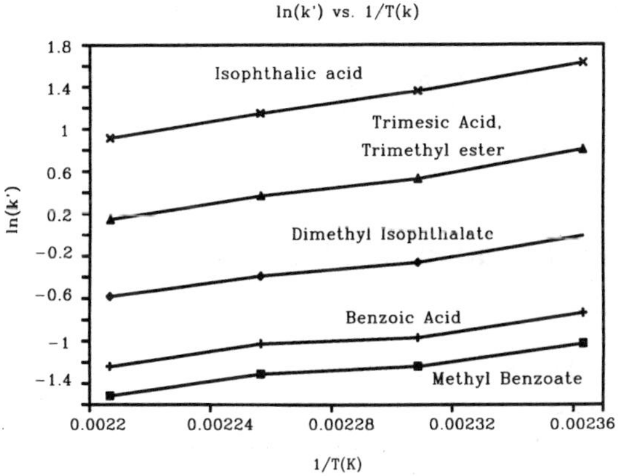

Figure 1
Variation of ln k' with $1/T$ for methyl benzoate, benzoic acid, dimethyl isophthalate, trimethyl trimesate, and isophthalic acid eluting from a DB-5 column with a mobile phase of CCl_2F_2 at a density of 0.3 g ml^{-1}.

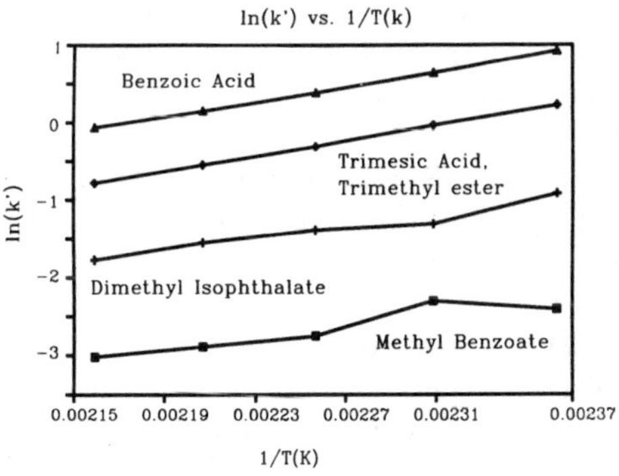

Figure 2
Variation of ln k' with $1/T$ for methyl benzoate, dimethyl isophthalate, trimethyl trimesate, and benzoic acid eluting from a Carbowax column with a mobile phase of CCl_2F_2 at a density of 0.3 g ml^{-1}.

Further investigations are in progress to obtain ΔH°_T values for several of these solutes using CO_2 and other fluorocarbon mobile phases on DB-5 and Carbowax as the stationary phase. Results of these studies will be reported at a later date. At this early stage, we can report that the increased polarity of CCl_2F_2 allows some polar molecules to be eluted from capillary SFC columns more rapidly than when CO_2, at the same density is used as the mobile phase. The use of the Hildebrand solubility parameter as a measure of the eluent strength of polar supercritical fluids does not appear to have much merit in forecasting retention behavior of polar comlounds in capillary SFC.

5.4 FT-IR Identification of Eluites with a Polar Mobile Phase

From the above discussion, it can be seen that polar mobile phases are potentially useful for capillary SFC separation of polar analytes. After the separation of mixtures of such compounds, it is important that techniques for the rapid, sensitive identification of each eluite be available. Mass spectrometry (MS) has been successfully applied to the characterization of components eluting from capillary SFC columns using CO_2 as the mobile phase [14,15]. For large molecules of the type for which SFC is the most relevant separation technique, electron ionization mass spectra are exceedingly difficult to interpret, and rarely yield much more structural information than the molecular weight unless reference spectra of all potential components are available. It is important to develop other identification techniques which will yield complementary data to the mass spectrum if detailed structural information is to become available. One such technique is Fourier transform infrared (FT-IR) spectrometry.

Two techniques for SFC/FT-IR measurements have been described. The first involves simply passing the effluent from the column through a high-pressure flow cell. This approach has been implemented by several groups [16–19], and recently Nicolet Analytical Instruments Corporation has announced the commercial introduction of an SFC/FT-IR interface based on this principle.

The principal limitation of this approach is the severe reduction of energy at wavelengths where the mobile phase has strong absorption bands. Of all fluids which have been used for capillary SFC, only xenon is completely free of absorption across the entire mid-infrared spectrum [20]. The spectrum of CO_2 at pressures only slightly in excess of P_c has several wide transmission windows, and only the two infrared active fundamentals, centered near 670 and 2350 cm^{-1}, and one combination band near 3500 cm^{-1}, are totally absorbing at the pathlengths typically required for SFC/FT-IR (about 5 mm). As the pressure of CO_2 is increased, however, several other bands start to increase in intensity and can present an annoying and often critical interference at pressures in excess of 200 atm.

The use of flow cells has been investigated for SFC/FT-IR measurements when $CO_2 : CH_3OH$ mixtures are used as the mobile phase [19,21]. The absorbance spectrum of 5% methanol in CO_2 at 204 atm in a 1-mm pathlength cell (compensated for absorption by the CO_2) is shown in **Figure 3**. This spectrum indicates that there would be less than 1% transmittance at all wavenumbers below 1600 cm^{-1}, and between 3700 and 2750 cm^{-1}, *i. e.* across almost the entire useful mid-infrared spectrum, if a cell with a 5-mm pathlength were used. Absorption by CO_2 further reduces the accessible spectral range. Similar considerations hold for capillary SFC/FT-IR measurements made with CCl_2F_2 mobile phase. In fact over most of the fingerprint region below 1400 cm^{-1}, CCl_2F_2 is opaque (transmittance < 1%) even with a pathlength as short as 100 μm.

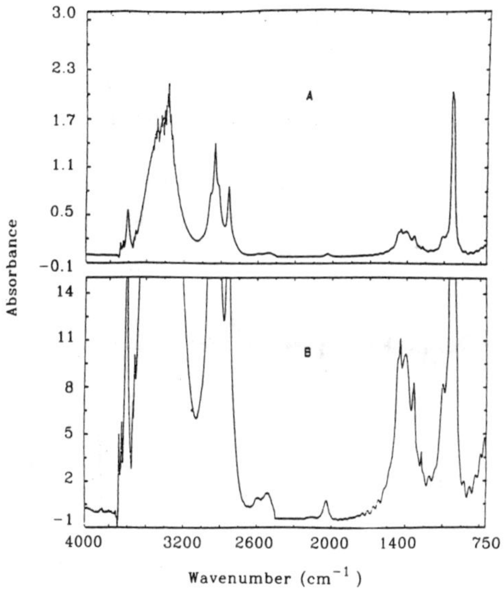

Figure 3
(A) Infrared absorbance spectrum of 5% methanol in CO_2 measured at 3000 psi and ambient temperature with a pathlength of 150 μm; all absorption features due to CO_2 have been compensated by spectral substraction. (B) Calculated absorbance spectrum of 5% methanol in CO_2 at 3000 psi for a pathlength of 5 mm. Note that the absorbance over most of the wavenumber region below 1600 cm^{-1} exeeds 2.0.

The second approach for SFC/FT-IR involves eliminating the mobile phase prior to the measurement of the spectrum of each eluite. In this case, a moving substrate is held about 100 μm below the end of the pressure restrictor. The mobile phase evaporates immediately on emerging from the restrictor while each eluite is deposited on the substrate as a spot of small diameter (typically $\approx$ 150 μm). By reducing the diameter in which the analytes are contained in this manner, their effective pathlength, and hence their absorbance is increased. It is necessary, however, to mount the substrate in an FT-IR microscope in order to get a high radiation flux through the samples, and hence to measure their spectra with a high signal-to-noise ratio.

An interface based on this principle has been described by us [22–24], but to date has only been tested for separations effected using CO_2 as the mobile phase. There is, of course, no reason why it should not be used with any SFC mobile phase, since all common supercritical fluids are gaseous at STP and will evaporate immediately on emerging from the restrictor. A simplified SFC/FT-IR interface was tested using a mixture of quinones separated by capillary SFC with CCl_2F_2 as the mobile phase, and found to work satisfactorily. As an example of its performance, the spectrum of 25 ng of 1,8-dihydroxyanthraquinone separated on a DB-5 column using CCl_2F_2 at a density of 0.3 g ml^{-1} is shown in **Figure 4**. Identifiable spectra have been obtained from components injected at a quantity of ≈ 1 ng when CO_2 was used as the mobile phase, and a minimum identifiable quantity of the same magnitude is confidently predicted when CCl_2F_2 and other fluorocarbons are used.

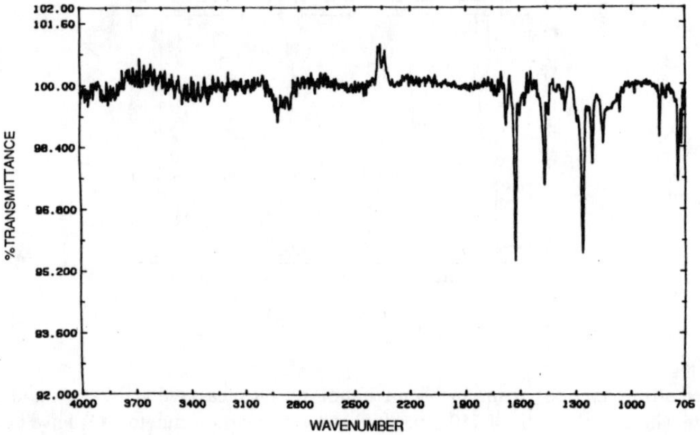

Figure 4
SFC/FT-IR spectrum of 25 ng of 1,8-dihydroxyanthraquinone after elution from a DB-5 column using CCl_2F_2 at 0.3 g ml^{-1}.

References

[1] *M. Novotny, S. R. Springston, P. A. Peaden, J. C. Fjeldsted,* and *M. L. Lee*, Anal. Chem. **53** (1981) 407 A–414 A.

[2] *B. A. Jones, K. E. Markides, J. S. Bradshaw,* and *M. L. Lee*, Chromatography Forum **1** (1) (1986) 38–44.

[3] *J. C. Giddings, M. N. Myers, L. McLaren,* and *R. A. Keller*, Science **159** (1968) 67–73.
[4] *J. H. Hildebrand* and *R. L. Scott*, "Regular Solutions", Prentice-Hall, Englewood, NJ (1962).
[5] *B. W. Wright, H. T. Kalinoski,* and *R. D. Smith*, Anal. Chem. **57** (1985) 2823–2829.
[6] *J. C. Fjeldsted* and *M. L. Lee,* Anal. Chem. **56** (1984) 619 A–628 A.
[7] *J. M. Levy* and *W. M. Ritchey*, J. Chromatogr. Sci **24** (1986) 242–248.
[8] *B. W. Wright* and *R. D. Smith*, J. Chromatogr. **355** (1986) 357–373.
[9] *D. J. J. Fraser* and *P. R. Griffiths*, unpublished results (1987).
[10] *G. L. Pariente* and *P. R. Griffiths*, unpublished results (1987).
[11] *G. L. Pariente, S. L. Pentoney, P. R. Griffiths,* and *K. H. Shafer*, Anal. Chem. **59** (1987) 808–813.
[12] *B. I. Lee* and *M. G. Kesler*, AIChE J **21** (1975) 349–354.
[13] *C. R. Yonker* and *R. D. Smith*, J. Chromatogr. **351** (1986) 211–218.
[14] *R. D. Smith* and *H. R. Udseth,* Anal. Chem. **59** (1987) 13–22.
[15] *H. T. Kalinoski, H. R. Udseth, E. K. Chess,* and *R. D. Smith,* J. Chromatogr. **394** (1987) 3–14.
[16] *K. H. Shafer* and *P. R. Griffiths*, Anal. Chem. **55** (1983) 1939–1942.
[17] *S. V. Olesik, S. B. French,* and *M. Novotny,* Chromatographia **18** (1984) 489–495.
[18] *C. C. Johnson, J. W. Jordan, L. T. Taylor,* and *D. W. Vidrine,* Chromatographia **20** (1985) 717–723.
[19] *J. W. Jordan* and *L. T. Taylor*, J. Chromatogr. Sci. **24** (1986) 82–88.
[20] *S. B. French* and *M. Novotny*, Anal. Chem. **50** (1986) 164–166.
[21] *K. H. Shafer,* "The Hyphenation of Chromatography and Spectrometry", Ph. D. Dissertation, Ohio University, Athens, OH (1985).
[22] *S. L. Pentoney, K. H. Shafer,* and *P. R. Griffiths,* J. Chromatogr. Sci. **24** (1986) 230–235.
[23] *S. L. Pentoney, K. H. Shafer, P. R. Griffiths,* and *R. Fuoco,* HRC & CC **9** (1986) 168–171.
[24] *P. R. Griffiths, S. L. Pentoney, A. Giorgetti,* and *K. H. Shafer,* Anal. Chem. **58** (1986) 1349 A–1366 A.

6 Ion Mobility Spectrometry as a Detection Method for Supercritical Fluid Chromatography

Herbert H. Hill, Jr. and Mark Morrissey
Department of Chemistry, Washington State University, Pullman, Washington 99164–4630, USA

Herbert H. Hill, Jr. received a B. S. degree (1970) from Rhodes College, Memphis, Tennessee. Under the guidance of *Walter Aue* he received a M. S. degree (1973) from the University of Missouri, Columbia and a Ph. D. (1975) degree from Dalhousie University, Halifax, Nova Scotia, Canada. He carried out postdoctoral research (1975–76) at the University of Waterloo under the direction of *F. W. Karasek* before joining the faculty at Washington State University, Pullman, Washington in 1976. At WSU, he has held positions of Assistant Professor of Chemistry (1976–80), Associate Professor of Chemistry (1980–85), and Professor of Chemistry (1985–present). During 1983–84 he spent a year as guest professor of Chemistry at Kyoto University, Kyoto, Japan sponsored by a fellowship from the Japan Society for the Promotion of Science. *Professor Hill's* research interest is in trace organic analysis with emphasis on ambient pressure detection methods for chromatography. His current projects include research in supercritical fluid chromatography and ion mobility spectrometry. He has published over 70 papers in the area of chromatography and chromatographic detection methods.

Mark Morrissey received a B. S. degree (1982) from the University of Wisconsin, Oshkosh and is currently completing his Ph. D. degree (1988) with *Prof. Hill* at Washington State University. During his graduate career, he received the C. Glenn King award (1982) for graduate studies and the chemistry department's excellence in teaching award (1983). His research focuses on the investigation of ambient pressure ionization detectors for supercritical fluid chromatography. He has five publications in this area.

Summary:

Although the ion mobility detector is still in its infancy as a fully characterized detection system for supercritical fluid chromatography, initial investigations of the detector have proved promising. The unidirectional flow design enables the measurement of reliable and reproducible mobility data even when foreign gases are introduced into the ionization region. Its response to compounds which do not contain chromophores extends the range of detection beyond those compounds which can be monitored by the UV/VIS detector. The ability to capture on-the-fly low resolution ion mobility spectra of individual chromatographic peaks can also provide qualitative information. A non-selective mode of operation enables the detector to produce FID-like chromatograms of high molecular weight compounds. A tunable mode of operation permits the selective detection of compounds contained in complex mixtures. Lastly, its compatibility with a wide variety of supercritical fluids frees the chromatographer to select mobile phases on the basis of separation requirements rather than detector requirements.

Concentrate:

This chapter describes a universal and selective ambient pressure ionization detector for supercritical fluid chromatography which can be used with a variety of mobile phases. Based on the principles of ion mobility spectrometry, the detector has two modes of operation with detections limits in the sub-nanogram/second range for most organic compounds. Monitoring positive ions in the non-selective mode provides a response similar to that obtained with a flame ionization detector. Monitoring negative ions in the non-selective mode provides a response similar to that obtained with an electron capture detector. When selected ion mobilities are monitored, it responds as a compound selective or class selective detector. This ion

mobility detector (IMD) has been successfully demonstrated for use with pressure programming supercritical fluid chromatography when CO_2, CO_2/MeOH, or Freon is used as the mobile phase.

6.1 Introduction

This chapter describes the development of a universal and sensitive detection method for supercritical fluid chromatography which is compatible with both capillary and packed columns. As discussed in previous chapters, a major advantage of supercritical fluid chromatography is that it can be employed with a wide variety of sensitive detection methods which are traditionally used in either liquid or gas chromatographic systems. For quantitative detection, the most common detectors used with SFC are the UV/Vis absorption detector [1,2] and the flame ionization detector [3,4]. For qualitative detection, FTIR [5] and mass spectrometry [6] are being investigated. Other promising, but less universal detection methods include, fluorescence [7], nitrogen/phosphorus selective surface ionization [8], and flame photometric detection [9].

Unfortunately, each of these detection methods is limited to specific chromatographic conditions. For example, the UV/VIS detector is limited to compounds which contain chromophores with high extinction coefficients and must be chromatographed using mobile phases which do not absorb light in the wavelength range of detection. The flame ionization detector, a sensitive and universal detector, is limited to the use of mobile phases which do not contain ionizable species. Similarly, the FTIR must employ mobile phases which either do not absorb in the information rich region of the spectra or that can be easily separated from the compounds of interest followed by off line detection [10,11]. And mass spectrometry limits the molecular weight range that can be investigated.

What appears to be happening in supercritical fluid chromatography is that detector requirements rather than separation requirements are determining the conditions under which a chromatograph must be operated. Clearly, what is needed is a sensitive but simple detection method which can be used to collect both quantitative and qualitative information under a variety of chromatographic conditions. Freed from what are sometimes very stringent detector limitations, chromatographers can then focus on optimizing separation and yet be assured of adequate detection.

Principles of ion mobility spectrometry (IMS) have recently been applied to chromatographic detection [12]. With multiple modes of operation, non-selective and selective quantitative information and semi-qualitative information can be obtained for sub-nanogram quantities of many organic compounds. This chapter provides an introduction to ion mobility spectrometry with emphasis on the advantages of ion mobility detection after supercritical fluid chromatography.

6.2 Ion Mobility Spectrometry

Ion mobility spectrometry is an atmospheric pressure, gas phase ion separation technique which has been used only sparingly in analytical chemistry. A typical spectrometer, shown in **Figure 1**, consists of an ionization region and an ion transit region. In the ionization region, gas phase ions are produced by a variety of ambient pressure ionization processes [13]. The most commonly employed ion source and the one discussed in this chapter is the ^{63}Ni source which is similar to that used in the electron capture detector. Once gas phase ions are produced, they are directed through the mobility region by a uniform electric field. Accelerated by the electric field and decelerated by collisions with the atmosphere, the ions exhibit a constant velocity as they travel through the transit region. When instrumental parameters such as temperature, pressure, and electric field are held constant, each ion's velocity is a measure of its reduced mass, size, and shape. Although this atmospheric pressure ion separation technique cannot compete with mass spectrometry in terms of overall resolution, its separation power is sufficient to offer many practical applications for chemical analysis.

As a chromatographic detector, the spectrometer can be operated in three modes, providing informatively different data from the same sample. Using a Fourier transform method of operation [14, 15], rapid on-the-fly ion mobility spectra from high resolution capillary peaks can be obtained. If the spectrometer is not scanned but rather tuned to respond to all of the ions produced by the chromatographed components, quantitative non-selective detection can be obtained [15, 16]. If the detector is tuned to respond only to ions of a selected velocity, the spectrometer can provide a continuous selective tracing of the chromatography [15, 16]. In some cases, the spectrometer has been found to operate as a class selective detector for compounds which do not contain hetero atoms [17].

When evaluating the ion mobility detector (IMD) as a method for SFC, several questions must be answered: 1) What drift gas should be used? 2) Are ion mobility values constant? 3) Do compounds without chromophores respond? 4) Do high molecular weight compounds respond? 5) Can compounds be detected selectively? And 6) what mobile phases can be used? In the following sections, each of these questions will be discussed with answers demonstrating the potential of ion mobility spectrometry as a universal detection method for supercritical fluid chromatography.

6.3 The Ion Mobility Drift Gas

The ion mobility drift gas is that gas which is flowing in the detector through which the ions migrate. Under normal operating conditions, this gas is introduced

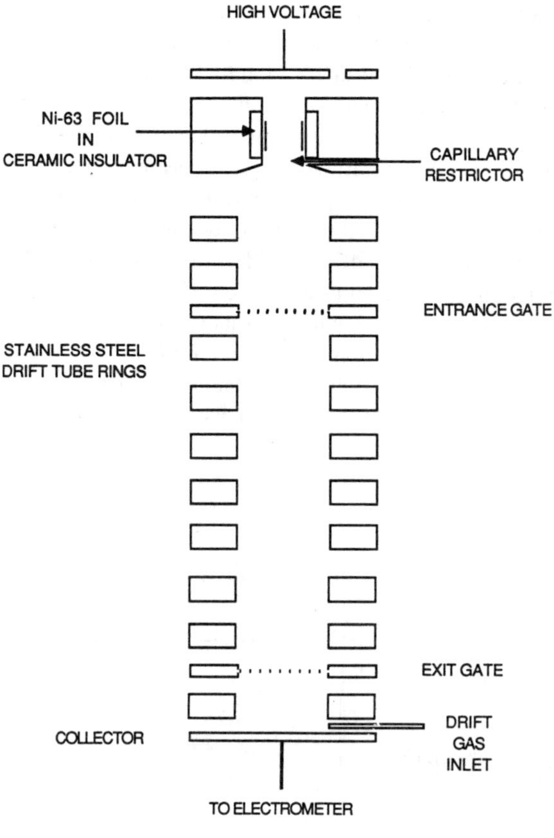

Figure 1
Schematic diagram of the ion mobility detector.

into the detector at a rate of about 400 to 600 ml/min in a direction which opposes the ion migration (see Figure 1). With SFC, when CO_2 is used as the mobile phase, it seems reasonable to ask "Can CO_2 also be used as the ion mobility drift gas?" While, in theory, it should be possible to use any inert gas as a drift gas, in practice, the most common drift gases are air and nitrogen. Thus, initial studies of SFC/IMD were performed with CO_2 as both the mobile phase and as the ion mobility drift gas [18].

These studies demonstrated that CO_2 could indeed be used as the drift gas and that ion mobility spectra obtained with CO_2 were, with two major differences, similar to those obtained with nitrogen. The two major differences observed in the mobility spectra were ion drift times and sensitivity. Although comparative spectra were similar, ion drift times were up to four times longer for CO_2 than with nitrogen. Because ions were in the drift region for a longer period of time and diffusion was greater, ion peaks broadened and the sensitivity of the detector was diminished. While these studies demonstrated that it was possible to use CO_2 as the drift gas, continued use of nitrogen or air appeared more desirable.

One potential problem with the use of nitrogen or air as the drift gas is contamination from the CO_2 that is introduced into the detector from the chromatograph. Mixing various amounts of CO_2 with nitrogen in the transit region, it was found that ion mobilities changed as a function of the amount of CO_2 in the drift gas [19]. Therefore, it is of paramount importance that mobile phase gases do not contaminate the ion transit region of the spectrometer. The unidirectional flow design, pictured in Figure 1, is required if the mobile phase from the chromatograph is to be efficiently swept out of the detector without contaminating the ion drift zone.

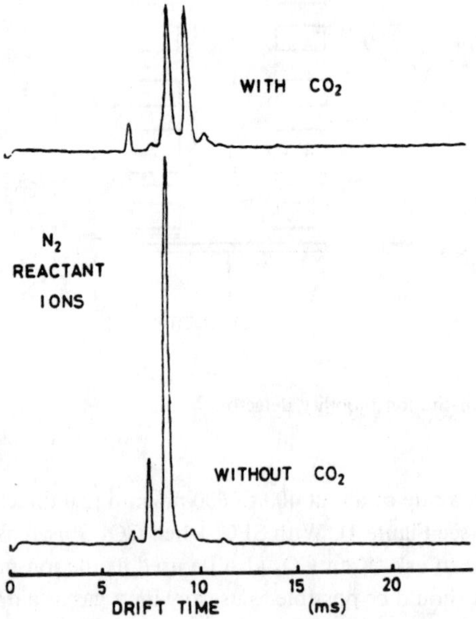

Figure 2
Effect of CO_2 on ion mobility reactant ions with standard ion mobility spectrometer. Using an ion mobility spectrometer with opposing gas flows CO_2 from an SFC column causes changes in the reactant ion spectrum. Lower tracing: Positive mode reactant ions in pure nitrogen. Upper tracing: Positive mode reactant ions when CO_2 is introduced into the ionization region of the spectrometer. (Ref. [18]).

Figure 2 demonstrates the effect that the CO_2 mobile phase has on the reactant ions in an ion mobility spectrometer which has not been modified for chromatographic detection. The lower tracing is an ion mobility spectrum obtained when only nitrogen is introduced into the detector while the upper tracing demonstrates that this spectrum changes significantly when the detector is interfaced to a supercritical fluid chromatograph. **Figure 3** shows the results of the same experiment performed with an ion mobility spectrometer which has been modified for chromatographic detection. Again, the upper tracings represent the results obtained for both positive and negative ions when the detector is not interfaced to a supercritical fluid chromatograph and the lower tracings show that the CO_2, being efficiently swept from the drift region by the unidirectional flow of the nitrogen drift gas, does not alter the spectrum.

6.4 Mobility Values

The final test to insure that ion mobilities do not change as a result of supercritical fluid introduction has been performed by comparing the mobilities of ions produced from compounds introduced into the detector by both gas chromatographic and supercritical fluid chromatographic methods [20]. **Table 1** reports the operating conditions used for both experiments. IMD parameters are representative of values used in general for IMD operation.

Table 1
Chromatographic and ion mobility conditions

	GC	SFC
Column	Fused silica	Fused silica
I. d.	250 μm	100 μm
Length	30 m	16 m
Stationary phase	DB-5	SPB-1
Coating thickness	0.25 μm	0.25 μm
Mobile phase	He	CO_2
Column temperature	250–280 °C	110 °C
Column pressure	1 atm	200–300 atm
IMD pressure	1 atm	1 atm
IMD temperature	300 °C	150 °C
IMD electric field	+ 240 V/cm	+ 240 V/cm
Drift length	7.2 cm	7.2 cm

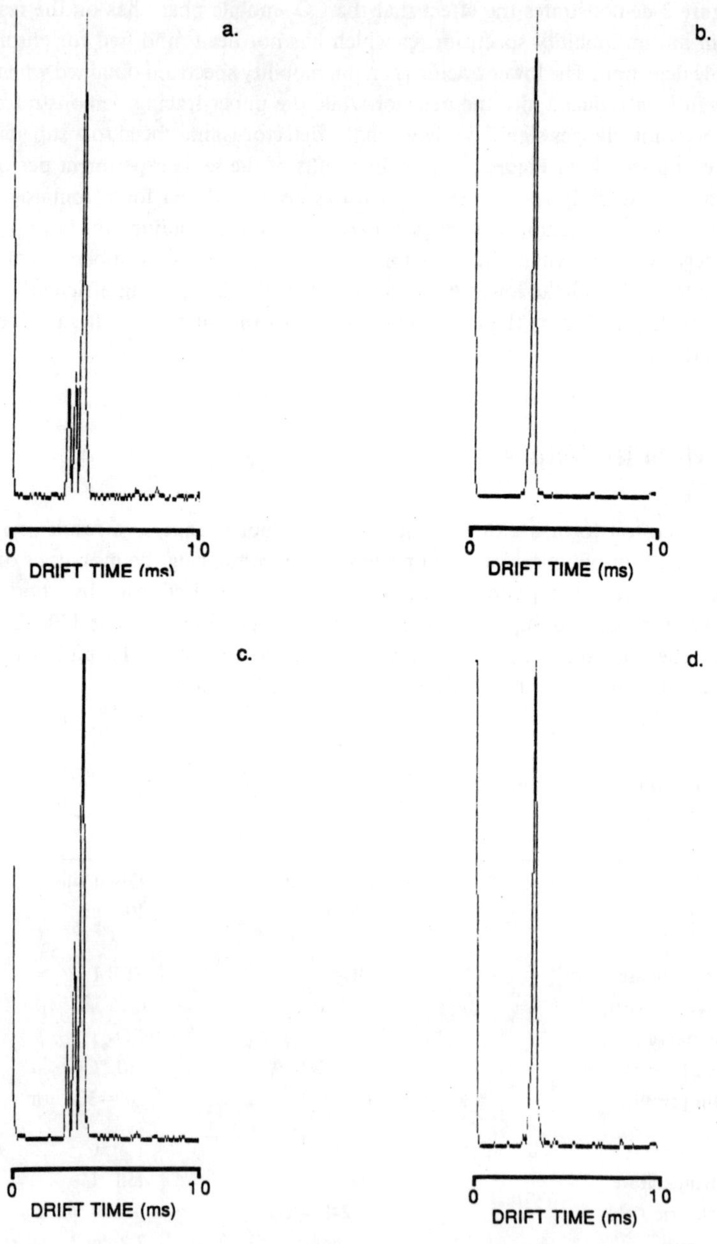
a.
0
10
DRIFT TIME (ms)
b.
0
10
DRIFT TIME (ms)
c.
0
10
DRIFT TIME (ms)
d.
0
10
DRIFT TIME (ms)

Under these conditions, testosterone, estrone, progesterone, diazepam, nitrazepam, and flurazepam were chromatographed and detected in both systems. The results of these studies are shown in **Table 2**. Note that although the observed drift times were considerably different for the two systems, so were the operating temperatures of the detector. With SFC the detector was operated at 150 °C while the IMD required an operating temperature of 300 °C for GC operation. Since, in ion mobility spectrometry mobilities are a function of diffusion, it is not surprising that drift times of the ions differed at different temperatures. For comparison of ion mobility data obtained under differing operating conditions, reduced mobility constants (K_0) are normally employed. These constants are a function of the electric field *(E)*, the drift distance *(d)*, the drift gas pressure in the tube *(P)*, and the drift tube temperature (*T*). They are calculated from measured ion drift times (*t*) using the following equation

$$K_0 = (d/Et)\ (273/T)\ (P/760)$$

Table 2
Comparative drift time (ms).

	GC (300 °C)	SFC (150 °C)
Steroids		
Testosterone	10.8	14.6
Estrone	10.3	14.3
Progesterone	11.3	15.4
Benzodiazophones		
Diazepam	10.2	14.0
Nitrazepam	10.2	13.8
Fluorazepam	12.4	16.4

◀ **Figure 3**
Comparison of reactant ions with and without CO_2 flow into the detector. (a) Positive mode reactant ions, CO_2 flow approximately 2 ml/min. (b) Negative mode reactant ions, CO_2 flow approximately 2 ml/min. (c) Positive mode reactant ions, no CO_2 flow. (d) Negative mode reactant ions, no CO_2 flow. All drift spectra recovered via Fourier transform from ion mobility interferograms, ion mobility detector 250 °C, 316 V/cm. (Ref. [17]).

As shown in **Table 3**, SFC and GC data compared favorably once the drift times were converted to mobility constants and corrected for temperature and pressure.

Table 3
Comparative mobility constants.

	GC (300 °C)	SFC (150 °C)
Steroids		
Testosterone	1.21	1.23
Estrone	1.27	1.25
Progesterone	1.16	1.16
Benzodiazophones		
Diazepam	1.29	1.28
Nitrazepam	1.29	1.29
Flurazepam	1.06	1.09

6.5 Detection of Compounds without Chromophores

One major advantage of the IMD as a detection method for SFC is that it responds to compounds which cannot be detected by standard UV/Vis methods. To demonstrate this point, a recent study has reported the serial arrangement of a UV/Vis absorption detector, typical of the type used for on column detection after microbore or packed capillary liquid chromatography, followed by introduction into an ion mobility spectrometer. In this study, methyl esters of fatty acids were separated by SFC and then monitored by both UV/Vis absorption and ion mobility spectrometry. As stated earlier, compounds used in initial investigations of the IMS were selected such that their IMS behavior could be evaluated by both GC and SFC. Although, methyl esters of fatty acids are routinely determined by capillary gas chromatography followed by flame ionization, they served as convenient non-chromophoric containing compounds for the initial evaluation of IMS after SFC. **Figures 4** and **5** demonstrate the sensitivity of IMS to compounds which could not be detected in the UV/Vis detector [18]. The upper tracing of Figure 4 follows the pressure program used to separate a mixture of methyl esters. The lower tracing represents the UV/Vis response of an SFC separation of this mixture under conditions which were sufficient to detect benzoate derivatives of a similar mixture of the fatty acids. The numbers above the lower tracing, mark the points in the chromatogram at which ion mobility spectra were obtained. Spectra for each of these scans are displayed in Figure 5. Scan 1 was taken prior to the injection and

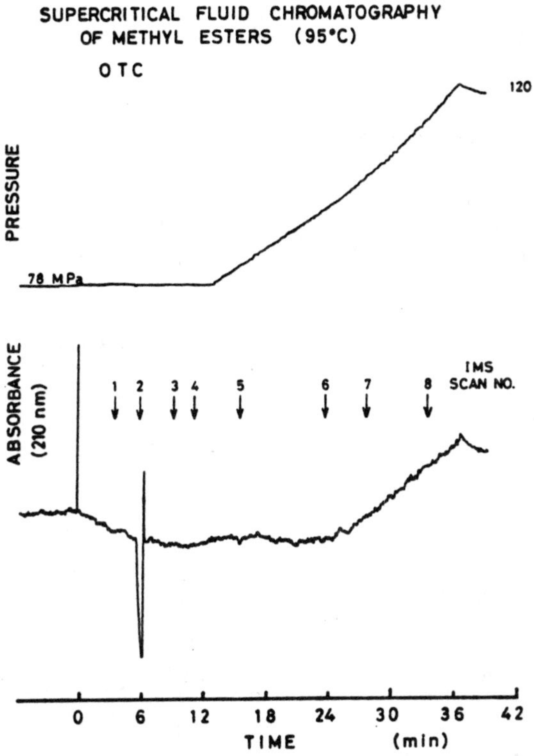

Figure 4
Pressure-programmed supercritical fluid chromatogram of methyl esters with UV detection. UV detector: on-column flow cell, 0.2 mm x 5 mm, internal volume of 150 ml, fixed wavelength of 210 nm, Jasco UVIDEC 100-II UV detector. Mobile phase: CO_2. Numbers mark position of IMS scans. (Ref. [18]).

represents the background (or reactant) ions present in the spectrometer. Scan 2 shows the ion mobility spectrum for the solvent, hexane. Scans 3, 4, 5 and 7 are individual ion mobility spectra obtained as each of the methyl esters eluted from the UV/VIS detector. Scan 6 was taken in the valley between peaks 5 and 7 and, as with the pre-injection scan, shows only the background ions.

Continuous chromatographic detection of this separation could be obtained by constantly monitoring a selected region of the ion mobility spectrum. If a wide region is monitored then chromatographic detection would be non-selective and if a narrow region is monitored then the detector's response would be selective. Ex-

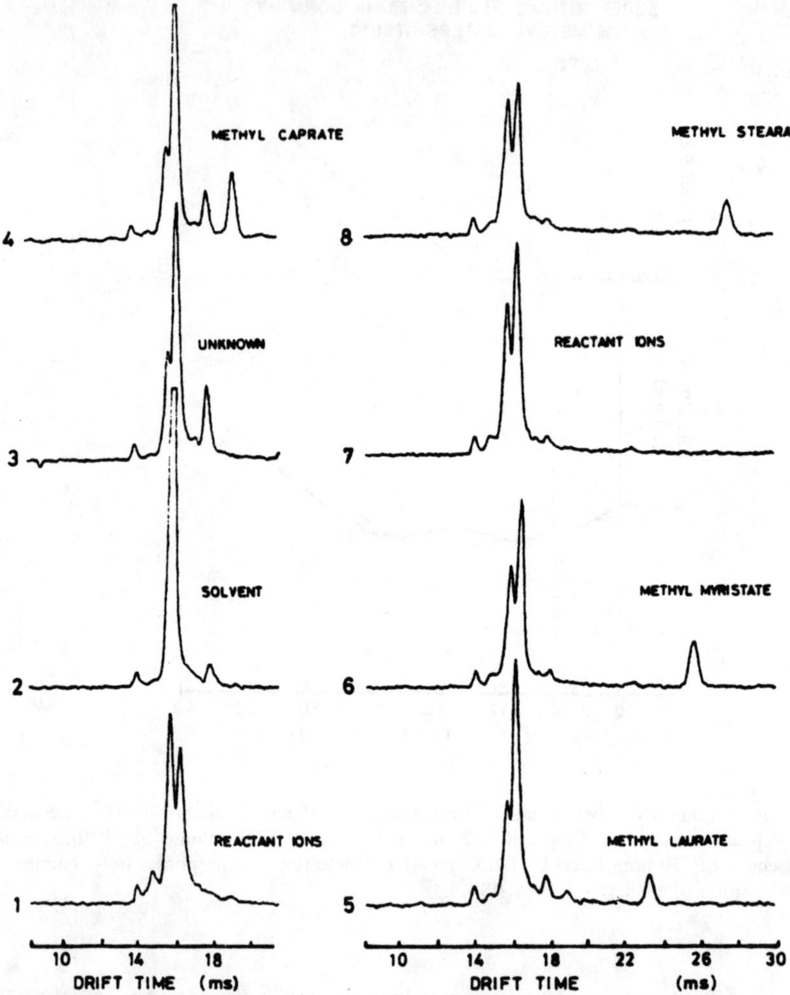

Figure 5
Ion mobility spectra of methyl esters after supercritical fluid chromatographic separation. Scan numbers correspond to those given in Figure 4. Compounds: 0.1 mg methyl cuprate, 0.1 mg of methyl laurate, 0.1 mg of methyl myristate, and 0.9 mg of methyl stearate. (Ref. [18]).

amples shown in the remainder of this chapter demonstrate the ion mobility spectrometer in this continuously monitoring mode of operation rather than in the scanning mode shown in Figure 5.

6.6 Molecular Weight Range

Prior to the development of SFC/IMD, the largest pure compound observed in ion mobility spectrometry had a molecular weight less than 500 amu. However, if the ion mobility detector is to be useful for SFC, it must be capable of detecting compounds in the 500 to 5,000 molecular weight range. **Figure 6** shows a supercritical fluid chromatographic separation of methyl silicone oligomers, average molecular weight of 2,000, with ion mobility detection [17]. The IMD was operated

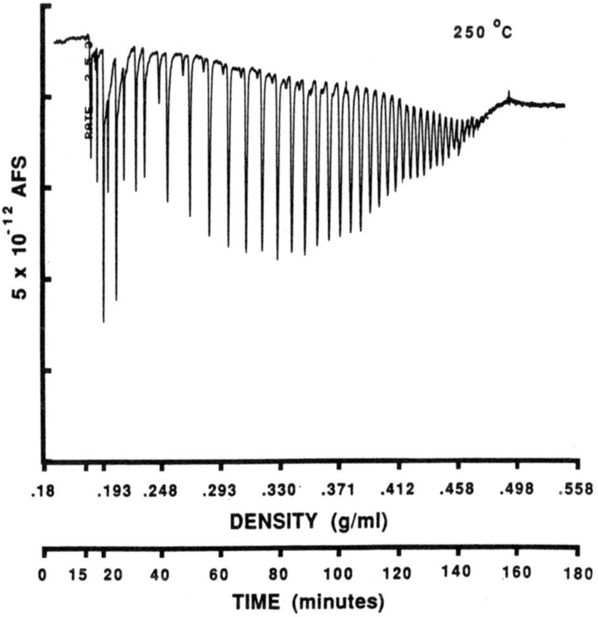

Figure 6
SFC/IMD of polymethylsiloxane: monitoring reactant ion depletion. CO_2 mobile phase. Density programmed from 0.18 g/ml to 0.71 g/ml at 0.003 g/ml/min after an initial 15 min constant density period. Temperature programmed from 100 to 140 °C at 0.25 °/min after an initial 20 min isothermal period. Density values shown in the figure are corrected for increasing temperature. Column J & W DB-1, 100 μm i. d. x 20 m, 0.4 μm film. IMD operated at 250 °C, + 316 V/cm. (Ref. [17]).

in a non-selective mode in which background current due to the reactant ions was continuously monitored. Under these conditions, individual oligomers with molecular weights greater than 3,000 amu were detected.

Detailed studies of the response factors as a function of molecular weight have not yet been completed but the patterns observed in Figure 6 and in other polymer separations suggest that the detector responds well to compounds in the SFC molecular weight range.

6.7 Selective Detection

A significant advantage of the IMD over the FID as a sensitive detection method for SFC is that not only can it provide non-selective responses as shown in Figure 6, but it can also be tuned to monitor a narrow range of the mobility spectrum. When this narrow window corresponds to the drift time of individual product ions, selective detection of specific compounds in a complex sample occurs.

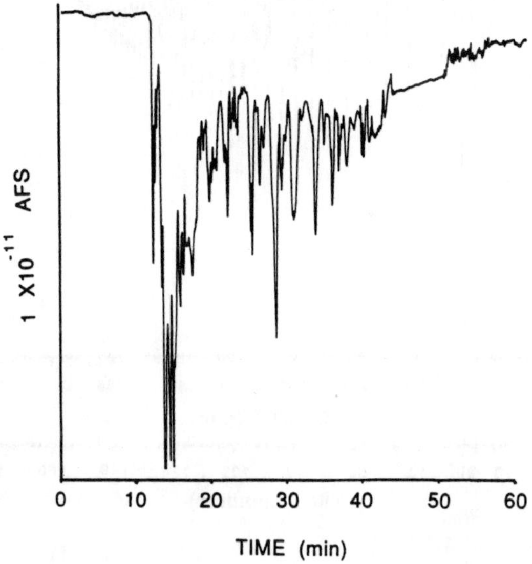

Figure 7
SFC/IMD of an underivittized soil extract. Monitoring reactant ion depletion. CO_2 mobile phase. Pressure programmed from 120 to 280 atm at 4 atm/min after an initial 10 min isobaric period. Oven operated isothermally at 110 °C. Column same as in Figure 5. IMD operated at 150 °C, −250 V/cm. (Ref. [17]).

Figure 7 is an SFC chromatogram of an underivatized soil extract containing 500 ppb of 2,4-dichlorophenoxyacetic acid (2,4-D) [17]. The standard procedure for the determination of 2,4-D in soil involves not only an extraction but also column clean-up and derivatization steps followed by high resolution capillary gas chromatography with electron capture detection [21,22]. **Figure 8** shows the selective detection by SFC/IMD of 2,4-D in the underivatized extract without column clean-up [17].

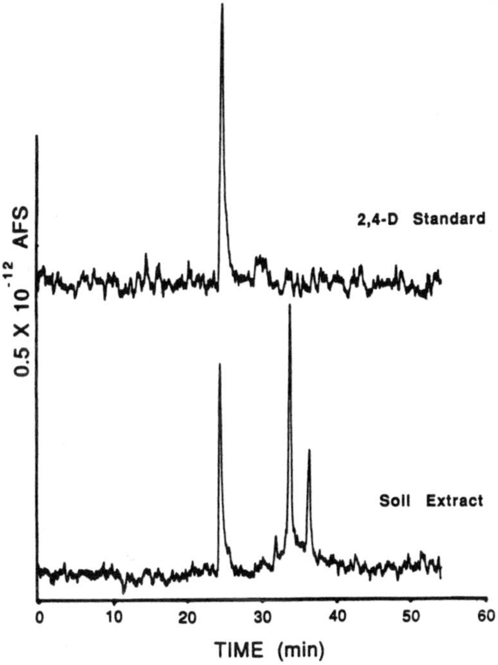

Figure 8
SFC/IMD of an underivatized soil extract. Selective product ion monitoring. Chromatographic and detection conditions are given in Figure 6. (Ref. [17]).

6.8 Mobile Phase Compatibility

The versatility of supercritical fluid chromatography has been limited because of the narrow range of mobile phases that can be practically employed to effect a given separation. To date, only relatively few phases are available for routine operation. To be sure, the physicochemical properties of phases and phase mixtures in relation to chromatographic compatibility narrow the choices which are available to the chromatographer. In addition, compatibility with the method of detection restricts the choice even more. With the FID, for example, the mobile phase is restricted to uncontaminated CO_2. In fact, herculean efforts are often required to clean the gas of ionizable impurities. **Figure 9** demonstrates the effect that ionizable modifiers have on the baseline of the FID as the column is pressure programmed from 100 atm to 300 atm [17]. For pure CO_2, the baseline remains stable throughout the program and permits the successful use of this detector. However, when 5% of methanol, methylene chloride, or hexane is added as a modifier, significant changes in the baseline occur as increasingly more of the ionizable modifier is introduced into the detector. In some cases, freons for example, flame ionization is so great and the combustion products are so caustic that the mobile phase cannot be used at all.

Because the ions produced from the low molecular weight compounds that are typically used as mobile phases behave similar to background ions normally found in the detector, the IMD can be used with a variety of mobile phases. In **Figure 10**,

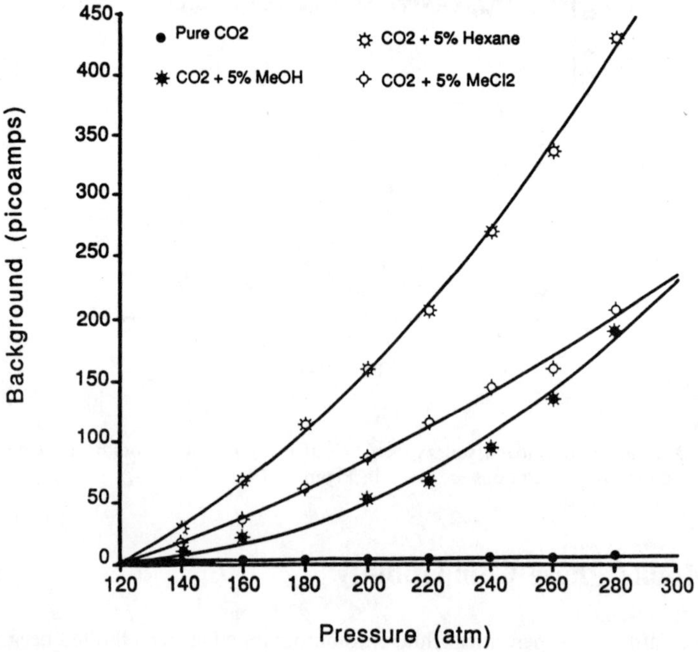

Figure 9
Effect of modifiers on the FID. Background current of the FID as a function of SFC system pressure. All initial background readings normalized to 0 picoamps. (a) Dry CO_2, (b) with 5% methanol. (c) CO_2 with 5% methylene chloride. (d) CO_2 with 5% hexane. (Ref. [17]).

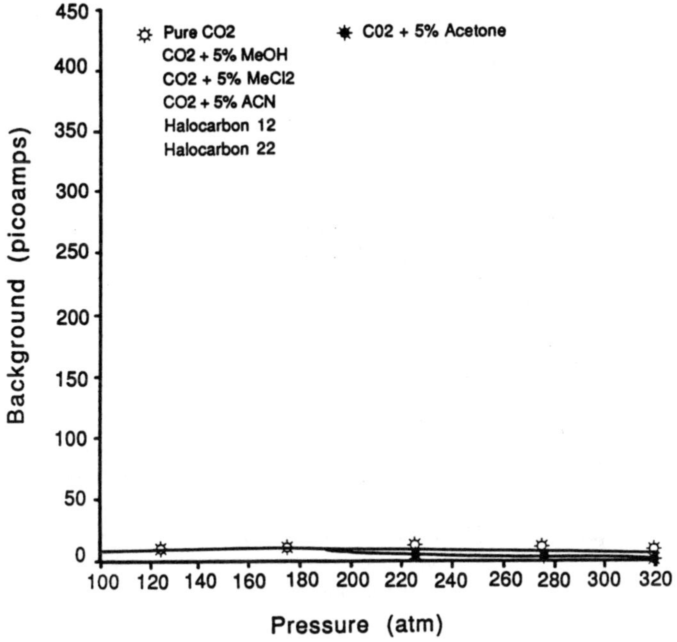

Figure 10
Effect of modifiers on the IMD. Background current of positive mode reactant ions as a function of SFC system pressure. (a) Dry CO_2. (b) CO_2 with 5% methanol. (c) CO_2 with 5% methylene chloride. (d) CO_2 with 5% acetonitrile. (e) CO_2 with 5% acetone. (f) Halocarbon 12, dichlorodifluoromethane. (g) Halocarbon 22, dichlorofluoromethane. (Ref. [17]).

the ion mobility detector was tuned to monitor only the background ions in the detector when a variety of fluids or fluid mixtures were used as the mobile phase [17]. As demonstrated in the figure, the background remained constant throughout the pressure programmed run.

It should be noted that response factors of the detector may change as a function of mobile phase and that complete detailed information on these effects are not currently available. **Figure 11** does, however, demonstrate that responses for methylsilicones remain similar to that obtained in pure CO_2 (Figure 6) when this polymer is separated using 5% methanol in CO_2 as the mobile phase [17].

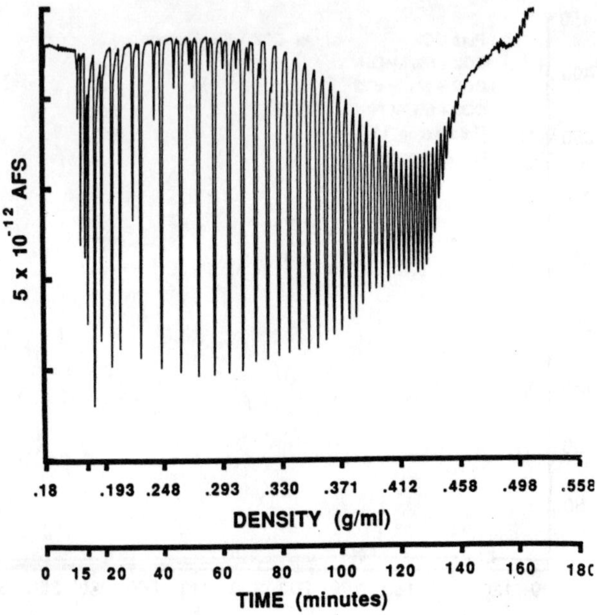

Figure 11
Effect of an organic modifier on IMD detection of polymethylsiloxane. CO_2 with 5% methanol mobile phase. All other conditions are given in Figure 5. (Ref. [17]).

References

[1] *J. C. Kuei, K. E. Markides,* and *M. L. Lee,* HRC & CC **10** (1987) 257–262.
[2] *A. L. Blilie* and *T. Greibokk* Anal. Chem. **57** (1985) 2239–2242.
[3] *D. W. Later, B. E. Richter, D. E. Knowles,* and *M. R. Andersen,* J. Chromatogr. Sci. **24** (1986) 249–253.
[4] *T. L. Chester, D. P. Innis,* and *G. D. Owens,* Anal. Chem. **57** (1985) 2243–2247.
[5] *K. H. Shafer* and *P. R. Griffiths,* Anal. Chem. **55** (1983) 1939–1942.
[6] *R. D. Smith, H. R. Udseth,* and *B. W. Wright,* J. Chromatogr. Sci. **23** (1985) 192–199.
[7] *P. A. Peaden, J. C. Fjeldsted, M. L. Lee, S. R. Springston,* and *M. Novotny,* Anal. Chem. **54** (1982) 1090–1093.
[8] *J. C. Fjeldsted, R. C. Kong,* and *M. L. Lee,* J. Chromatogr. **279** (1983) 449–455.
[9] *K. E. Markides, E. D. Lee, M. Bolick,* and *M. L. Lee,* Anal. Chem. **58** (1986) 740–743.
[10] *J. W. Hellgeth, J. W. Jordan, L. T. Taylor,* and *M. A. Khorassani,* J. Chromatogr. Sci. **24** (1986) 183–188.
[11] *D. Kuehl* and *P. R. Griffiths,* J. Chromatogr. Sci **17** (1979) 471–476.
[12] *M. A. Baim* and *H. H. Hill, Jr.,* Anal. Chem. **54** (1982) 38–43.

[13] *M. A. Baim* and *H. H. Hill, Jr.*, "Plasma Chromatography as a Chromatographic Detection Method" in Plasma Chromatography, *T. W. Carr* (ed.) Plenum Pub. Corp., New York and London (1984).
[14] *F. J. Knorr, R. L. Eatherton, W. F. Siems,* and *H. H. Hill, Jr.,* Anal. Chem. **57** (1985) 402–406.
[15] *R. L. Eatherton, W. F. Siems,* and *H. H. Hill, Jr.,* HRC & CC **9** (1986) 44–48.
[16] *R. L. Eatherton, M. A. Morrissey, W. F. Siems,* and *H. H. Hill, Jr.,* HRC & CC **9** (1986) 154–160.
[17] *M. A. Morrissey* and *H. H. Hill, Jr.,* unpublished data.
[18] *S. Rokushika, H. Hatano,* and *H. H. Hill, Jr.,* Anal. Chem. **59** (1987) 8–12.
[19] *S. Rokushika, H. Hatano,* and *H. H. Hill, Jr.,* Anal. Chem. **58** (1986) 361–365.
[20] *R. L. Eatherton, M. A. Morrissey,* and *H. H. Hill, Jr.,* unpublished data.
[21] *J. F. Thompson,* ed., "Analysis of Pesticide Residues in Human and Environmental Samples," U. S. Environmental Protection Agency, Research Triangle Park, NC (1977), Sections 10, 11.
[22] *P. A. Biondi* and *M. Cagnasso,* J. Chromatogr. **109** (1975) 389–394.

7 Elution Behavior of Nitrogen-Containing Compounds of Different Basicity from Various Stationary Phases Employing a Supercritical Mobile Phase

M. Ashraf-Khorassani and L. T. Taylor
Virginia Polytechnic Institute and State University, Department of Chemistry, Blacksburg, VA 24061-0212, USA

Larry T. Taylor joined the Virginia Tech faculty as an Assistant Professor in 1967. He spent two and one-half years as a National Institutes of Health Postdoctoral Fellow at the Ohio State University. A native of South Carolina, *Dr. Taylor* was promoted to Professor of Chemistry in 1978. During the summers of 1976 and 1984 he was a NASA-ASEE summer faculty fellow at the Langley Research Center, Hampton, VA. He received the Sporn Award for excellence in freshman teaching in 1977. He has authored more than 120 refereed technical publications plus numerous technical reports for the Department of Energy, Electric Power Research Institute, National Aeronautics and Space Administration, National Institutes of Health, Aluminum Company of America, Standard Oil Company of Ohio, *etc. Dr. Taylor* is currently the co-author of four patents in the areas of analytical and polymer chemistry. He currently serves as Associate Editor for the Journal of Chromatographic Science. He organized the "HPLC-FTIR" 1986 Summer Symposium for the ACS Division of Analytical Chemistry. *Professor Taylor's* research interests include the development and application of hyphenated (chromatography-spectroscopy) analytical techniques for the identification of polar material in complex matrices (*e. g.* GC-FTIR, SFC-FTIR, HPLC-FTIR).

Mehdi Ashraf-Khorassani is a native of Yazd, Iran. He received his B. S. in Chemistry and Biology from Jacksonville State University in 1982, and his M. S. in Electroanalytical Chemistry from the University of Southwestern Louisiana in 1984. In September 1984 he began his graduate studies toward the Doctor of Philosophy degree in the Department of Chemistry of Virginia Tech, under the direction of *Professor L. T. Taylor.* His primary research interests include chromatography, separation and identification of polar mutagens in coal-derived products and techniques for the analysis of biological samples.

Summary:

The separation via supercritical fluid chromatography of nitrogen-containing compounds which possess pK_a's ranging from −4 to + 11 has been investigated. Packed columns (4.6 mm i. d.) with the following stationary phases were employed: silica, octadecyl, propylamino and polystyrene-divinyl benzene. Isobaric mobile phases were 100% CO_2 and 99% CO_2/1% CH_3OH. Without modifer the range of basicities which could be eluted increased in the order: silica < PRP < C_{18} = NH_2. Chromatographic peak shapes were much better for NH_2 than C_{18}. Modifier promoted the elution of all components except aliphatic amines (pK_a > 10) from the four columns. Separation behavior using silica-based columns can be understood by noting the fraction of accessible silanol sites on each stationary phase. The propylamino unlike the C_{18} phase is believed to afford deactivation of adjacent silanol sites via internal hydrogen bonding. This interaction is believed to also inhibit the reaction of CO_2 with the propylamino bonded phases.

Concentrate:

Nitrogen-containing compounds with basicities ranging from pK_a = −3 to pK_a = + 7 have been separated via SFC employing a propylamino bonded silica packed column and CO_2 as the mobile phase.

7.1 Introduction

Nitrogen-containing compounds are common to pharmaceuticals, food products, pesticides, propellant stabilizers, drugs, tobacco smoke, automobile pollution, urban suspended particulate matter, coal-derived products, *etc.* Since most of these materials are complex matrices, a number of chromatographic procedures for isolating specific nitrogen-containing compounds from the mixture have been reported such as thin layer chromatography [1], paper chromatography [2], electrophoresis [3], cation exchange chromatography [4], gas chromatography [5], and high performance liquid chromatography [6]. Each technique has its limitations, especially as the need arises to separate more complicated mixtures of a broader polarity and molecular weight range. For example, capillary GC yields reasonable results with both flame ionization and nitrogen-selective detectors, but derivatization of the analyte is often necessary because many nitrogen-containing compounds are either not volatile or are thermally labile at the temperature required for GC analysis [8]. HPLC can deal with a broader range of compounds, but the analysis time and chromatographic resolution become limiting factors given the current state-of-the-art.

Supercritical fluid chromatography (SFC) currently offers an attractive alternative for the separation of these polar, thermally labile or nonvolatile components [9]. The greater diffusivity and lower viscosity afforded by a supercritical fluid relative to a liquid has been demonstrated to yield faster, more efficient separations; while "liquid-like" densities enable many thermally-inaccessible components to be solubilized. Both packed and capillary columns which utilize selective as well as universal detectors have been employed [10].

One of the first reports [11, 12] regarding SFC of polar compounds involved the separation of amino acids and sugars using either supercritical ammonia or CO_2 as the mobile phase. It was concluded that NH_3 was the better medium for the separation of polar materials, while CO_2 was better suited for non-polar compounds. Later *Lauer et al.* [13] reported the separation of caffeine, theophilline, and nicotine from a PRP-1 packed analytical scale column with supercritical NH_3. With a bonded silica stationary phase many of these polar materials can be separated with modified supercritical CO_2. For example, *Henion et al.* [14] employed CO_2 with 10–20% methanol as a polar modifier to elute a variety of polar drugs such as codeine, cocaine, and caffeine from a propylamino column. The separation of mixtures of xanthines, carbamates, sulfonamides, steroids, and ergot alkaloids has been reported using CO_2 and 10–15% methanol or methoxyethanol modified CO_2 [15]. Again, a propylamino bonded phase column was used. Carbamate pesticides [16, 17] and nitrated polycyclic aromatic hydrocarbons [18, 19] have also been investigated.

No exhaustive study has appeared specifically regarding the optimized separation of basic nitrogen-containing mixtures via SFC. *Jordan* [20] showed the separation of five basic nitrogen-containing compounds on propylamino bonded phase silica employing CO_2 with a Fourier transform infrared spectrometer as an on-line high information detector. Emphasis in this study was placed upon detection of the material rather than optimization of the separation. More recently, a direct comparison of HPLC *versus* SFC for the separation of primarily azaarenes again with a propylamino column and CO_2 has been communicated [21]. The effectiveness of other packed column stationary phases with this class of compounds is not known. The following report addresses this issue. Nitrogen-containing compounds of differing basicity have been investigated employing silica, propylamino (NH_2), octadecyl (C_{18}), and divinylbenzene cross-linked polystyrene (PRP-1) stationary phases with both supercritical fluid CO_2 and methanol modified CO_2.

7.2 Experimental

A Hewlett Packard (Avondale, PA) Model 1082B liquid chromatograph modified for use with supercritical fluids was utilized for studies with analytical scale (4.6 mm, i.d.) columns. A variable wavelength UV detector (Hewlett Packard Model 79875) equipped with a high pressure 1 cm pathlength (8 μl internal volume) flow cell was employed. Samples were injected onto the packed column via a Rheodyne Model 7125 six-port injection valve having a 10 μl loop.

Model nitrogen-containing compounds employed in this study were obtained from either Aldrich Chemical Company or Fisher Scientific Co. Each sample was dissolved in HPLC grade methylene chloride (Fisher Scientific Co.) prior to introduction onto the column. The concentration of samples per component ranged from 2 μg/μl for early eluting components to 3 μg/μl for late eluting components. Typically, 2 μl of each solution were applied to the columns. A flow rate of 2 ml/min was employed throughout the study. Conditions of pressure (5000 psi) and temperature (50 °C) were kept constant in order to provide a uniform evaluation of each column. Silica and bonded phase (propylamino and octadecyl) silica-based columns (25 cm length, 4.6 mm i.d., 5 μm particle diameter) were obtained from Alltech. The percent coverage provided by the octadecyl phase was 12% according to the manufacturer and was stated to also be endcapped. The percentage coverage of the propylamino phase was unknown and the column was not endcapped. The PRP-1 column (15 cm length, 4.6 mm i.d., 5 μm particle diameter) was purchased from Hamilton. Carbon dioxide and methanol-modified CO_2 were obtained from Scott Specialty Gases.

7.3 Results and Discussion

The object of this research was to separate nitrogen-containing compounds of different basicity on packed columns (4.6 mm i. d.) containing a variety of stationary phases via both supercritical CO_2 and methanol-modified supercritical CO_2 mobile phases. Different factors in addition to basicity such as steric hindrance of the nitrogen lone pair and solubility of the compounds in supercritical CO_2 were studied in relation to the retention mechanism. In this fashion we hoped to match the chemical and physical properties of certain compounds with the optimized stationary phase in order to achieve the best separation. A discussion of our results with four stationary phases (silica, C_{18}, NH_2, and PRP-1) follows. The model nitrogen-containing compounds employed along with their aqueous pK_a values and solubility in subcritical CO_2 (where known) are listed in **Table 1**.

Table 1

Compound	pK_a	Solubility[a) (25 °C, 950 psi)
Pyrrole	−3.8	NA[b)
Quinoxaline	0.56	NA
Pyrazine	0.65	NA
Diphenylamine	0.79	1%
Aniline	4.63	3%
N-Methylaniline	4.83	20%
N,N-Dimethylaniline	5.15	M
Pyridine	5.25	M
2,6-Lutidine	NA	NA
N,N-Dimethylbenzylamine	> 10	NA
Triethylamine	11.01	NA

a) *A. W. Francis*, J. Phys. Chem. **58** (1954) 1099.
b) NA = Data not available, M = miscible.
c) Data (aqueous) taken from "CRC Handbook of Chemistry and Physics", 60th Ed., CRC Press, Inc., Boca Raton, FL (1980).

7.3.1 PRP-1 Column

All separations were conducted at a fixed pressure (5000 psi) and temperature (50 °C). Two supercritical mobile phases were commonly employed (100% CO_2 and 99% CO_2/1% CH_3OH). Under these conditions some of our model compounds failed to elute from the PRP-1 column. Only materials with pK_a less than 1.0 could be removed with 100% CO_2 provided no steric hindrance of the nitrogen lone pair of electrons was evident. For example, quinoxaline, pyrazine, diphenylamine, and pyrrole with pK_a values of less than 1.0 were eluted from the column, while pyridine and aniline with higher pK_a's were not eluted. On the other hand, *N, N*-dimethylaniline possessing a pK_a = 5.15 but a sterically hindered nitrogen lone pair is eluted with a k' = 1.63. In like manner 2,6-lutidine is also eluted (**Table 2**). The single methyl group afforded by *N*-methylaniline apparently does not present sufficient steric hindrance since this compound does not elute (**Figure 1**). Incorporation of 1% methanol into the supercritical CO_2 resulted in the elution of all components except aliphatic amines (pK_a > 10). In addition, chromatographic peak shapes were much improved relative to the situation with 100% CO_2. A decrease in selectivity of the column was however, noted, but this no doubt could have been minimized had we chosen to change pressure-temperature parameters in going from CO_2 to modified CO_2.

Table 2
PRP-1 column.

Compound	Capacity factor (k') 100% CO_2	99% CO_2 + 1% CH_3OH
Pyrrole	1.54	0.59
Quinoxaline	5.77	1.26
Pyrazine	5.76	0.11
Diphenylamine	7.26	5.25
Aniline	NE[a)]	1.84
N-Methylaniline	NE	1.30
N,N-Dimethylaniline	1.63	0,98
Pyridine	NE	0.45
2,6-Lutidine	2.25	0.48
N,N-Dimethylbenzylamine	NE	NE
Triethylamine	NE	NE

a) NE = Not eluted: 5000 psi, 50 °C.

The role of the methanol modifier with the PRP-1 column is puzzling. It may provide (although it is unexpected at this low modifier level) a greater solvating power to the mobile phase. The stationary phase is expected to exhibit no reactive groups which might otherwise be deactivated by the methanol. However, if impurities reside

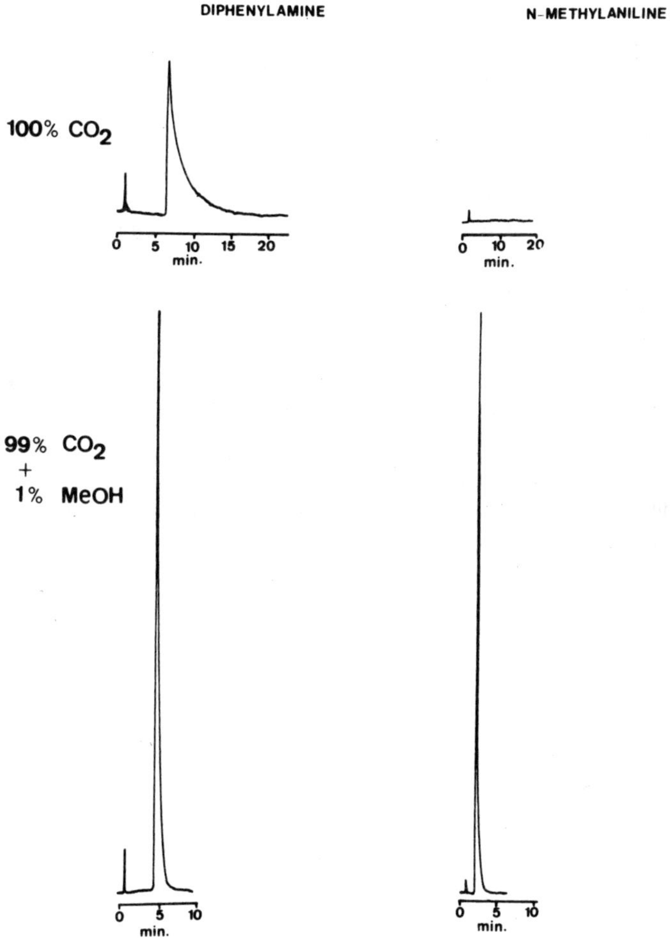

Figure 1
Elution behavior of diphenylamine and *N*-methylaniline from PRP-1 column with 100% supercritical CO_2 and methanol-modified supercritical CO_2 (99/1.0). UV detection, 254 nm.

on the polystyrene packing material, methanol may serve to deactivate these contaminants. Clearly a better understanding of the PRP-1 stationary phase is required before an accurate assessment of this phenomenon can be made.

7.3.2 Silica Column

Basic nitrogen-containing compounds such as those shown in Table 1 can be adequately separated on silica via HPLC by varying the polartiy of the mobile phase. However, with non-polar supercritical CO_2 as the mobile phase competition with the analyte for the stationary phase is not very great. Consequently, most of the materials did not elute from the silica column. Specifically, only pyrrole with $pK_a < 0$ and highly sterically hindered diphenylamine and *N,N*-dimethylaniline eluted with 100% CO_2. Addition of 1% CH_2OH to the mobile phase had a marked effect on the elution, **Table 3**. In this situation only the highly basic aliphatic amines failed to be mobilized. It again appears doubtful that the small amount of CH_3OH added to the mobile phase could promote a significant increase in solubility. On the other hand, there could be sufficient CH_3OH present to interact with and effectively neutralize (deactivate) the numerous accessible, free silanol sites via hydrogen bonding with either the silanol oxygen **(A)** or silanol hydrogen **(B)**. In other words, the silica column would become less active in the presence of polar modifier and the moderately basic analytes would be less retained. The improvement in chromatographic peak shape with methanol-modified CO_2 also supports this contention, **Figure 2**.

Table 3
Silica column.

Compound	Capacity factor (k') 100% CO_2	 99% CO_2 + 1% CH_3OH
Pyrrole	0.75	0.50
Quinoxaline	NE[a)]	3.55
Pyrazine	NE	2.23
Diphenylamine	1.12	0.42
Aniline	NE	1.32
N-Methylaniline	NE	0.66
N,N-Dimethylaniline	3.50	0.41
Pyridine	NE	4.57
2,6-Lutidine	NE	6.25
N,N-Dimethylbenzylamine	NE	NE
Triethylamine	NE	NE

a) NE = Not eluted: 5000 psi, 50 °C

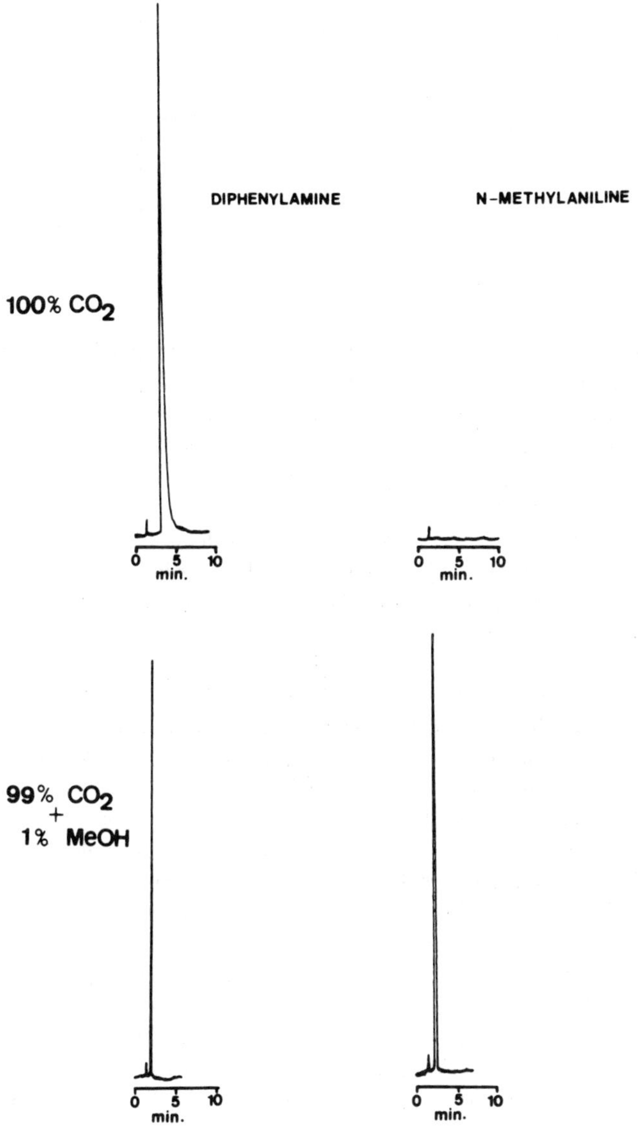

Figure 2
Elution behavior of diphenylamine and *N*-methylaniline from a silica column with 100% supercritical CO_2 and methanol-modified supercritical CO_2 (99/1.0). UV detection, 254 nm.

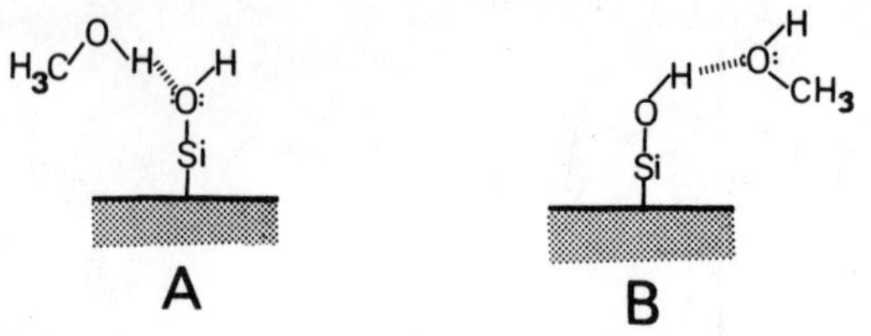

7.3.3 Octadecyl Column

Employing 100% CO_2, the C_{18} column permitted all materials with $pK_a < 6$ to elute regardless of steric hindrance. However, the chromatographic peak shapes suggest in most cases an irreversible interaction between analyte and stationary phase. Most compounds (*e. g.* pyrazine, aniline, *N*-methylaniline, 2,6-lutidine) eluted with a low efficiency and large tailing except for the weakly basic or sterically hindered analytes (*e. g.* pyrrole, dimethylaniline, and diphenylamine). It was originally believed that the nitrogen-containing compounds would exhibit little affinity for the C_{18} phase since the material had been endcapped following 12% bonded phase coverage. In other words, any analyte-column interaction would be via the accessible (to analyte) unreacted silanol groups rather than via partitioning with the octadecyl bonded phase.

Since a fair degree of selectivity was realized with C_{18} and 100% CO_2 as the mobile phase, we believe that such an interaction to be a valid one. The accessibility of these uncapped silanols may be poor since all peaks are broadened, suggesting that the interaction is not reversible. Evidence for this point was obtained when 2,6-lutidine (**Figure 3**) and pyridine were individually injected three times each onto the column. Retention times for each peak decreased as the number of injections to the column were increased. This decrease in retention and large peak tailing indicates that the stationary phase may be becoming saturated by irreversible adsorption of basic lutidine (or pyridine) to the column (*i. e.* free silanol sites). Addition of 2% CH_3OH to the poorly solvating CO_2 phase caused the adsorbed lutidine (no sample injected) to be spontaneously stripped from the column.

As expected chromatographic separation with 1% CH_3OH modified CO_2 eluted all components except the aliphatic amines when the mixture was injected on a rejuvenated column. Efficiency was increased, however, selectivity was extremely poor relative to the situation when 100% supercritical CO_2 was used (**Table 4**).

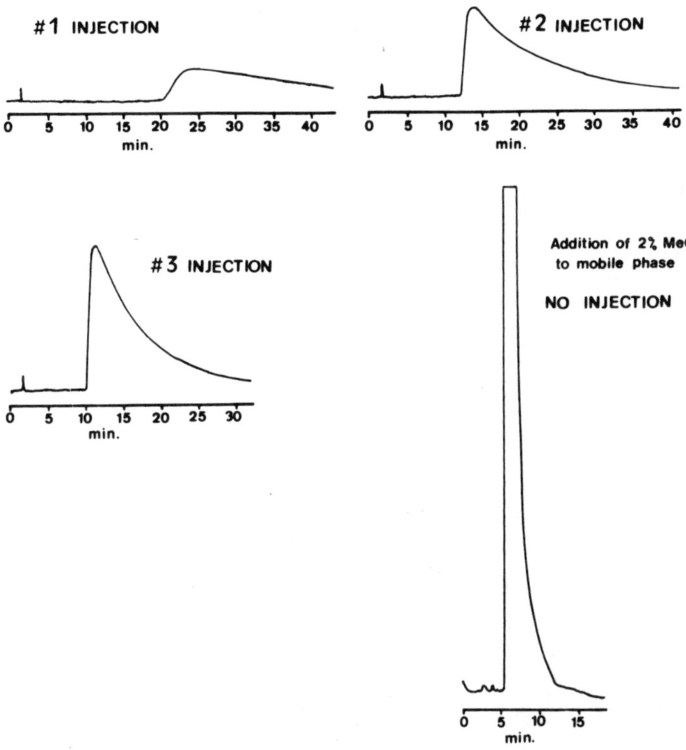

Figure 3
Three consecutive injections of 2,6-lutidine onto octadecyl column with 100% supercritical CO_2 as the mobile phase followed by addition of 2% CH_3OH to the mobile phase UV detection, 254 nm.

7.3.4 Propylamino Column

This column was expected to result in reasonable separations since previous work had shown that nitrogen-containing compounds could be easily chromatographed under similar conditions [21]. These previous results also suggested that the mechanism for retention depended upon the analyte's pK_a, the degree of nitrogen lone pair steric hindrance and the analyte's solubility in the supercritical medium. All the model compounds studied here except the aliphatic amines were eluted with 100% CO_2 under the standard pressure (5000 psi) and temperature (50 °C) conditions (**Table 5**). The relatively weak solvating power of supercritical CO_2 was again

Table 4
Octadecyl column.

Compound	Capacity factor (k') 100% CO_2	99% CO_2 + 1% CH_3OH
Pyrrole	0.09	0.02
Quinoxaline	4.39	0.37
Pyrazine	9.76	0.17
Diphenylamine	0.40	0.37
Aniline	1.80	0.16
N-Methylaniline	1.50	0.17
N,N-Dimethylaniline	0.58	0.22
Pyridine	8.36	1.07
2,6-Lutidine	10.7	3.22
N,N-Dimethylbenzylamine	NE[a]	NE
Triethylamine	NE	NE

[a] NE = Not eluted: 5000 psi, 50 °C.

Table 5
Amino column.

Compound	Capacity factor (k') 100% CO_2	99% CO_2 + 1% CH_3OH
Pyrrole	1.36	0.83
Quinoxaline	2.40	0.29
Pyrazine	1.14	0.16
Diphenylamine	1.88	0.74
Aniline	5.00	1.12
N-Methylaniline	2.29	0.34
N,N-Dimethylaniline	0.45	0.16
Pyridine	4.13	0.45
2,6-Lutidine	3.42	0.54
N,N-Dimethylbenzylamine	NE[a)]	NE
Triethylamine	NE	NE

[a)] NE = Not eluted: 5000 psi, 50 °C.

apparently unable to disrupt the strong interaction between the two aliphatic amines and residual silanols even through the amines which were injected are highly hindered. Peak shape for the eluted compounds was much improved over analogous

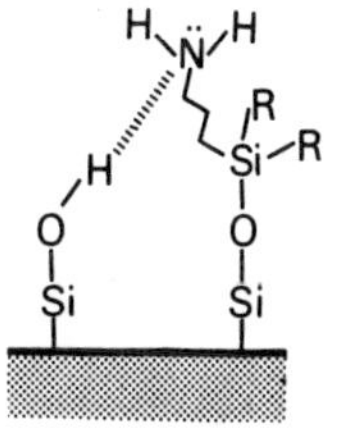

C

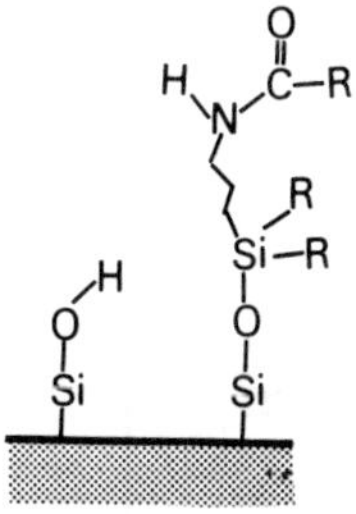

D

results with any of the other stationary phases (**Figure 4**). In comparison with the C_{18} column, the same number of components were eluted but they were eluted as much sharper peaks from the amino column even though this stationary phase had not been endcapped. Improved column performance relative to C_{18} may be attributed to several factors: (1) greater bonded phase coverage for the propylamino, (2) internal hydrogen bonding of propylamino groups with accessible silanol hydroxyls, **C**, and (3) derivatization of the propylamino group by reaction with CO_2 to yield a "urea-like" functionality, **D**. While the relative extent of silica coverage by the propylamino and octadecyl bonded phases is not known, we do not believe that the difference is significant. Deactivation of residual silanols via interaction with the polar bonded phase, however, appears to be an attractive rationale for the improved chromatographic performance of propylamino with 100% CO_2 relative to octadecyl which does not contain a polar bonded phase functionality. This, likewise, would be the case for silica which contains no bonded phase. Reaction of the stationary phase with CO_2, thereby converting the amino bonded phase to an amide or urea phase, structure **D**, cannot be discounted, especially in light of the fact that primary unhindered aliphatic amines readily react with CO_2 in a non-polar solvent. In order to determine whether the stationary phase had been reacted with CO_2, both diffuse reflectance Fourier transform infrared spectrometry and X-ray photoelectron spectroscopy were applied to the packing material both before and after exposure to supercritical CO_2. Apparently there was no reaction since no spectrometric changes were noted between these two materials. Structure **C** which would afford the nitrogen less nucleophilicity no doubt could account for this surprising lack of reactivity with CO_2.

The use of 1% methanol in the supercritical CO_2 failed to elute the aliphatic amines from the amino column. A modifier effect was, nevertheless, observed for the other components as evidenced by the higher capacity factors obtained with 100% CO_2. In contrast to the results obtained with the octadecyl phase and

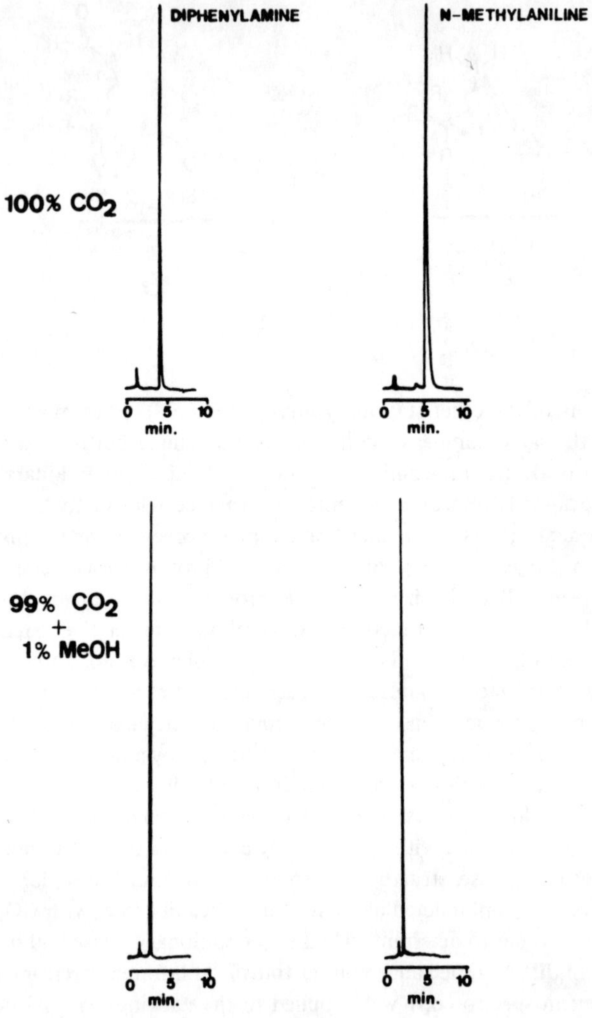

Figure 4
Elution behavior of diphenylamine and *N*-methylaniline from a propylamino column with 100% supercritical CO_2 and methanol-modified supercritical CO_2 (99/1.0). UV detection, 254 nm.

modifier where all analytes essentially eluted together, selectivity was in part retained with the propylamino phase and modifier.

It therefore, appears that with the silica and C_{18} stationary phases the primary interaction of the basic nitrogen-containing compounds (100% CO_2) is via activated silanol groups. The amino phase gives the best results with 100% CO_2 relative to silica and C_{18} since the number of free silanol groups is reduced by both silanol derivatization and internal hydrogen bonding through the propylamino group, Structure **C**.

With 1% methanol the best selectivity is also found with the propylamino bonded phase. On the other hand, very poor selectivity (*i. e.* poorer than silica) is observed with C_{18} wherein most components eluted with $k' < 0.4$. The reason for selectivity being retained may lie in the fact that a slightly different mechanism of interaction is introduced with the propylamino phase in the presence of methanol. For example, 1% methanol no doubt deactivates a large fraction of silanols in both the C_{18} and NH_2 case. The mode of analyte interaction in the C_{18} case would have to be via the deactivated silanols, **E**. For the amino situation, **F**, the analyte may choose to interact with either the polar bonded phase or the deactivated silanols.

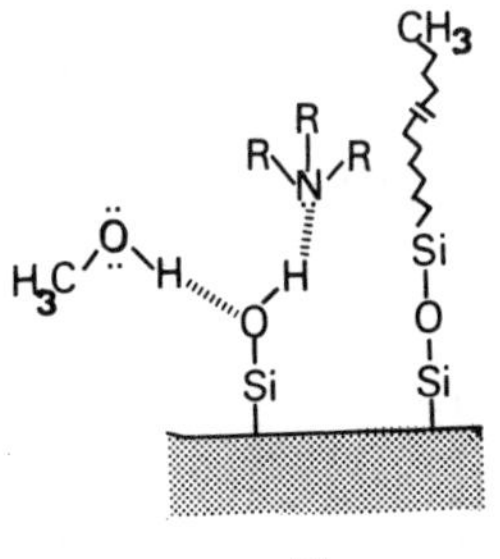

E

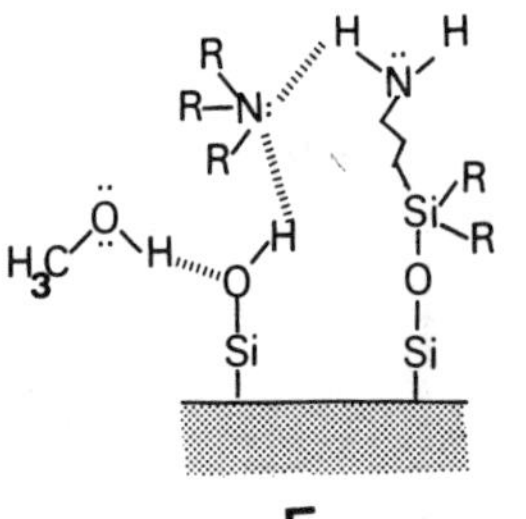

F

The superiority of amino relative to silica and C_{18} stationary phases is apparent even with 1% methanol as revealed in **Figure 5** where the separation of various methylated pyridines is shown for the four stationary phases. Only the PRP-1 and propylamino columns eluted all four compounds. 2,4,6-Collidine has a $pK_a = 7.43$ which explains its failure to be eluted from the more activated silica and octadecyl columns. Pyridine and the lutidines with $pK_a < 6$ were easily removed from these columns with 1% methanol-modified CO_2. A measure of the goodness of propylamino relative to silica, octadecyl, and PRP-1 is further shown in **Figure 6**, where the separation of isomeric benzoquinolines is compared. Methanol-modified CO_2 is required for elution of all four components from silica, C_{18}, and PRP-1; whereas, the same materials are removed from the propylamino column with only

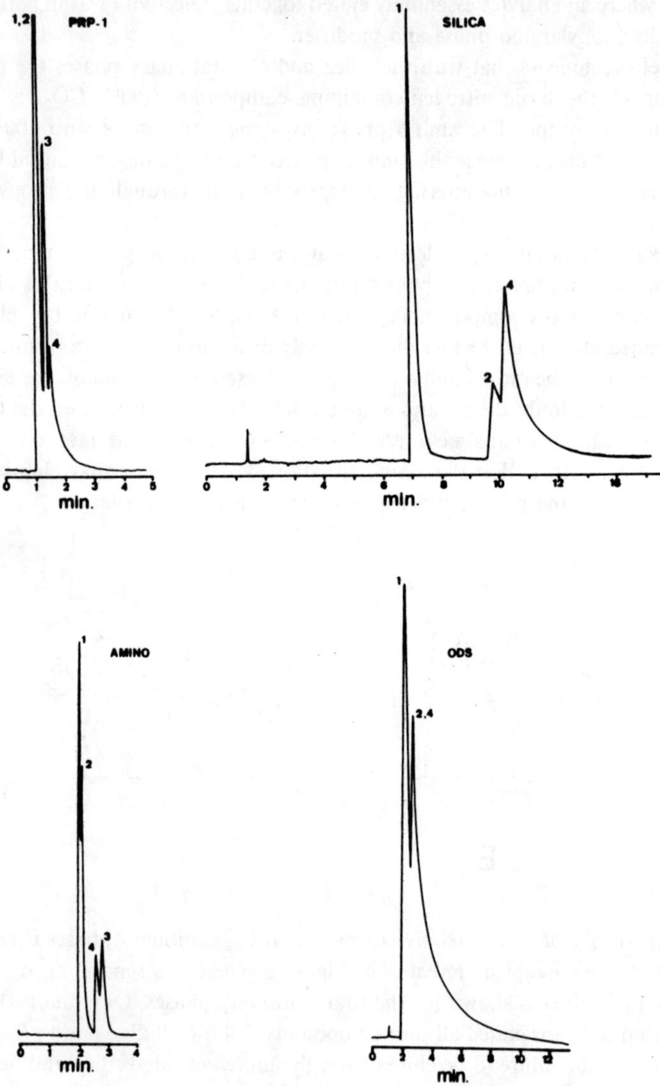

Figure 5
Isobaric separation (5000 psi, 50 °C, 99% CO_2/1% CH_3OH) of methylated pyridines on various stationary phases. UV detection, 254 nm. 1 = pyridine, 2 = 2,6-lutidine, 3 = 2,4,6-collidine, 4 = 3,5-lutidine.

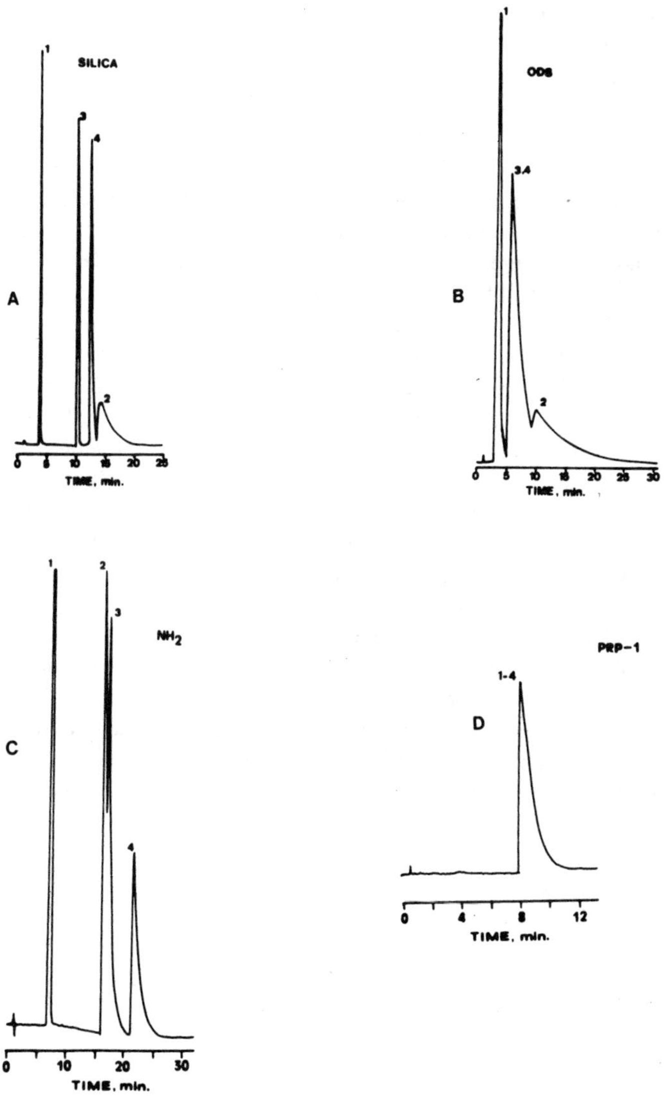

Figure 6
Isobaric separation (5000 psi, 55 °C) of benzoquinoline isomers on various stationary phases. Silica (99% CO_2/1% CH_3OH); ODS or C_{18} (99% CO_2/1% CH_3OH); PRP-1 (99% CO_2/1% CH_3OH); Amino (100% CO_2). UV detection, 254 nm. 1 = 7,8-benzoquinoline; 2 = 2,3-benzoquinoline; 3 = 3,4-benzoquinoline; 4 = 5,6-benzoquinoline.

100% supercritical CO_2. The more activated silica column demonstrated longer retention times than the C_{18} column as would have been expected.

Even though the propylamino functionality is expected to deactivate a fraction of the silanols, the presence of methanol in the supercritical CO_2 has a profound modifier effect. A number of azaarenes were successively separated **(Figure 7)** with 100% Co_2, 99.9% CO_2/0.1% CH_3OH, 99.8% CO_2/0.2% CH_3OH, and 98% CO_2/2% CH_3OH. The reduction in retention time and improvement in chromatographic peak shape with increasing percentage of methanol are dramatic. Holding pressure and temperature constant, it would appear that 0.2% CH_3OH is the optimized modifier content for this column and these compounds. It is also interesting to note that the order of elution is not maintained with increasing modifier content (*i. e.* 7,8-benzoquinoline, quinoline and quinazoline interchange).

In summary, nitrogen-containing compounds with basicities ranging from $pK_a = -3$ to $pK_a = +7$ have been separated via SFC employing packed columns. The presence of methanol as a modifier to the supercritical CO_2 promotes elution and improved peak shapes in each case. While the modifier effect with the PRP-1 column is not readily explained, it would appear that for silica based columns (silica, C_{18}, and NH_2) the methanol deactivates silanol sites via hydrogen bonding. The propylamino phase yielded the best results as judged by both the basicities that could be eluted without resorting to modifier and the absence of peak tailing. Partitioning of the analytes with the octadecyl phase seems to be minimal. Interaction with accessible silanol sites appears to dominate the separation mechanism in the absence of modifier. In the presence of modifier, partitioning with the propylamino phase is suggested. Clearly SFC of polar materials on silica-based stationary phases would be enhanced and greater selectivity would be achieved if these residual silanol sites could be either permanently capped or made to be less accessible.

Acknowledgment

The financial assistance provided by Department of Energy Grant DE-FG22-81PC40799 is greatly appreciated. Many informative and useful discussions regarding this work with *Dennis Gere* are gratefully acknowledged.

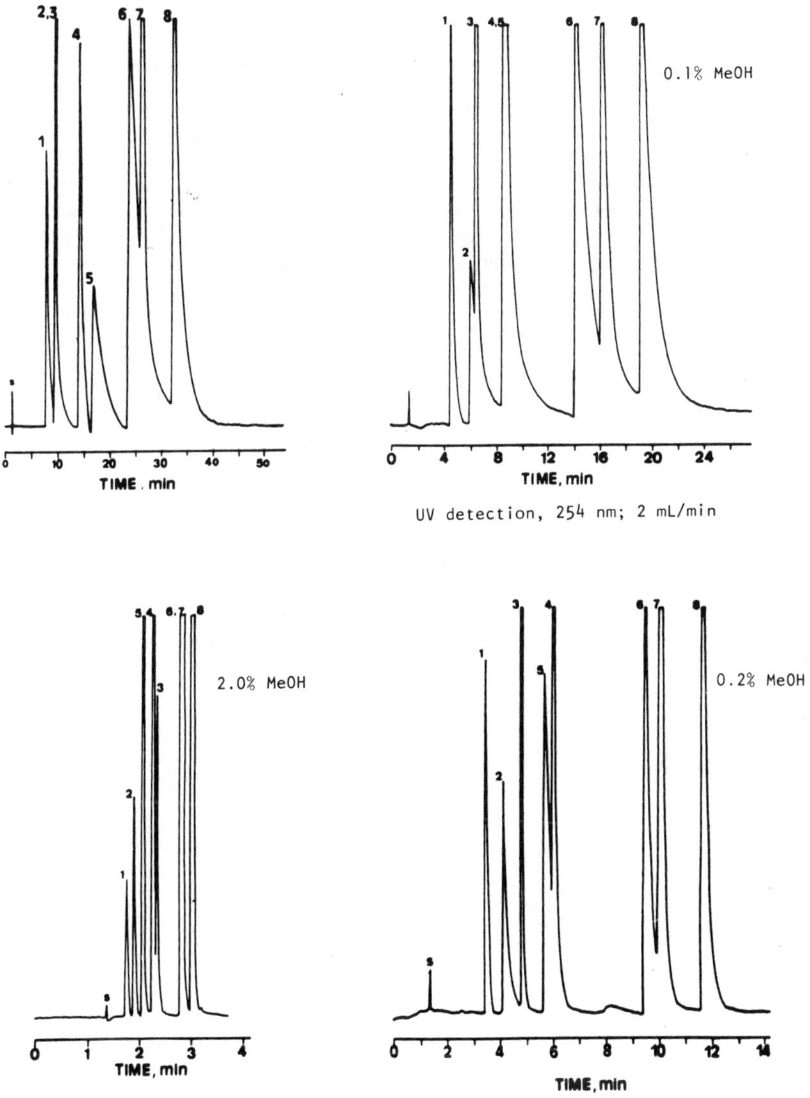

Figure 7
Isobaric separation (5000 psi, 50 °C) of azaarenes on propylamino column (250 mm x 4.6 mm, i. d., 5 μm) with variable amounts of modifier. UV detection, 254 nm, 2 ml/min, 1 = quinoxaline, 2 = pyridine, 3 = 7,8-benzoquinoline, 4 = quinoline, 5 = quinazoline, 6 = acridine, 7 = phenanthridine, 8 = 5,6-benzoquinoline.

References

[1] *A. S. Nalang, D. R. Choudhury,* and *A. J. Richards,* J. Chromatogr. Sci. **22** (1982) 235.
[2] *E. Sawicki,* and *J. D. Pfaff,* Anal. Chim. Acta. **32** (1965) 521–534.
[3] *E. Wawicki, M. Guyer,* and *C. R. Engel,* J. Chromatogr. **30** (1967) 522–527.
[4] *N. L. Holy* and *T. Y. Lin,* J. Liq. Chromatogr. **2** (1979) 687–698.
[5] *C. H. Ho, B. R. Clark, M. R. Guerin, C. Y. Ma,* and *T. K. Rao,* Fuel **24** (1979) 282–288.
[6] *P. G. Amateis* and *L. T. Taylor,* Anal. Chem. **56** (1984) 966.
[7] *S. W. Cooper* and *R. K. M. Jayanty,* J. Chromatogr. Sci. **24** (1986) 204–209.
[8] *J. A. Sweetman* and *F. W. Karasek,* J. Chromatogr. **247** (1982) 245–254.
[9] *C. M. White* and *R. K. Houck,* HRC & CC **9** (1986) 4–17.
[10] *M. L. Lee* and *K. E. Markides,* HRC & CC **9** (1986) 652–656.
[11] *L. McLaren, M. N. Myers,* and *J. C. Gidding,* Science **159** (1968) 197–199.
[12] *L. McLaren, M. N. Myers, R. A. Keller,* and *J. C. Gidding,* Science **162** (1968) 67–73.
[13] *H. H. Lauer, D. McManigill,* and *R. D. Board,* Anal. Chem. **55** (1983) 1370–1375.
[14] *J. B. Crowther* and *J. D. Henion,* Anal. Chem. **57** (1985) 2711–2716.
[15] *A. J. Berry, D. E. Games,* and *J. R. Perkins,* J. Chromatogr. **363** (1986) 147–158.
[16] *B. W. Wright* and *R. D. Smith,* HRC & CC **8** (1985) 8–11.
[17] *B. W. Wright* and *R. D. Smith,* HRC & CC **9** (1986) 73–77.
[18] *A. L. Blilie* and *T. Greibrokk,* Anal. Chem. **57** (1985) 2239–2242.
[19] *J. M. Levy* and *W. M. Ritchey,* J. Chromatogr. Sci. **24** (1986) 242–248.
[20] *J. W. Jordan* and *L. T. Taylor,* J. Chromatogr. Sci. **24** (1986) 82–88.
[21] *M. Ashraf-Khorassani* and *L. T. Taylor,* J. Chromatogr. Sci. **26** (1988) 331–336.

8 Applications of SFC in the Petroleum Industry

Herbert E. Schwartz
Applied Biosystems, Brownlee Labs Division, 2045 Martin Avenue, Santa Clara, CA 95050, USA
Joseph M. Levy and John P. Guzowski
BP America, Research and Development, 4440 Warrensville Center Road, Cleveland, OH 44128, USA

Joseph Michael Levy is a native of Cleveland, Ohio. He received his B. S. in Chemistry, Magna Cum Laude, from Cleveland State University in 1980, two M. S. degrees in Analytical and Organic Chemistry from Case Western Reserve University in 1983, and a Ph. D. in Physical Chemistry from Case Western Reserve University in 1986. *Joe* obtained his degrees while working full-time as a senior research associate at BP America Research and Development. He has been there since 1980 working in the Chromatographic Problems Analytical section. His principal research interests have been in the areas of high resolution gas chromatography and supercritical fluid chromatography. *Joe* has developed a number of multidimensional capillary GC methods for the characterization and determination of selected components in refinery, environmental, and reaction product streams. He has also used packed and capillary column SFC for petroleum related applications. This includes considerable research on the uses of modifiers in SFC, the development and use of on-line multidimensional SFC/capillary GC, and the use of supercritical fluid extraction coupled with SFC and GC. *Joe* is a member of the American Chemical Society and has published extensively in the area of analytical chemistry.

John P. Guzowski, Jr. is a Cleveland native and received his B. S. in Chemistry from John Carroll University in 1985. Since that time, he has been working at BP America Research and Development as a Research Chemist in the Chromatographic Problems Analytical Section. His research interests include high resolution gas and supercritical fluid chromatography as related to the petroleum industry and multidimensional chromatographic techniques. Other interests include the development of on-line chemical sensors and laboratory automation. *John* is a member of the local and national chapters of the American Chemical Society.

Herbert E. Schwartz is a native of Amsterdam, the Netherlands. He received a Masters degree in Chemistry in 1975 from the Free University, Amsterdam under the direction of *Prof. U. A. Th. Brinkman*. During 1976–1978 he worked as a research chemist for the Institute of Environmental Studies and as a technical editor for Excerpta Medica (Elsevier Science Division), a biomedical information system. In 1979 he moved to the U.S. where in 1983 he received a Ph. D. degree in Analytical Chemistry from Northeastern University, Boston under the direction of *Prof. B. L. Karger*. His doctoral thesis was on the use of microbore columns in liquid chromatography and on-line liquid chromatography/mass spectrometry. Since 1983 *Schwartz* has been employed by Applied Biosystems (formerly Brownlee Labs) in Santa Clara, California. He has authored *ca.* 20 papers in journals dealing HPLC, SFC, and environmental chemistry. Currently, he is product manager for SFC instrumentation at Applied Biosystems, Santa Clara division.

Summary:

One of the major application areas of SFC is in the petroleum industry. In this chapter, hydrocarbon group separations, the analysis of polynuclear aromatic hydrocarbons and simulated distillation by SFC are discussed as well as relevant

papers dealing with GC and HPLC. Although in many cases HPLC offers good selectivity for group separations, its main limitation has been, and still is, a lack of a universal detector which could provide equal response for individual components. SFC on the other hand can be easily combined with the flame ionization detector (without flow splitting) when packed microbore or capillary columns are used. SFC can be applied to group analysis of petroleum materials ranging from gasoline to crude oil residues and current approaches with CO_2 and SF_6 as mobile phases will be reviewed as well as a method based on FTIR detection. For group separations packed columns are most useful as high resolution is often not required while higher capacity and faster analysis time than with capillary columns are feasible. In some cases, column switching techniques are applied to further increase selectivity and/or to speed up analysis time.

Multidimensional techniques, *e.g.* on-line combination of SFC with GC, can provide detailed information on individual components by serving as a "fingerprint" technique. Combining low-resolution SFC with high-resolution GC is feasible without using a retention gap as is commonly done in HPLC/GC. Another area in which SFC traditionally has been applied is the analysis of polynuclear aromatic hydrocarbons (PAH). Examples on packed and capillary columns using pentane, CO_2 and ammonia as mobile phase in conjunction with FID and mass spectrometric detection will be discussed. The utility of modifiers (added to CO_2) coupled with UV detection will also be highlighted. Finally, recent examples in which SFC was applied to simulated distillation (SIMDIS) will be discussed. Compared to GC-SIMDIS where temperatures exceeding 400 °C are required for the elution of high boiling materials, much lower operating temperatures are used with SFC (typically 100–150 °C), thereby virtually eliminating the risk of sample decomposition. Examples of SFC-SIMDIS with packed as well as capillary columns are demonstrated.

Concentrate:

In this chapter, three petroleum related SFC applications areas are discussed: hydrocarbon group analysis, the analysis of PAH and simulated distillation. Examples showing the utility of FID detection for various group separations are given. In addition, multidimensional techniques combining low-resolution SFC with high-resolution GC are examined. Both packed and capillary columns can be used for PAH separations and simulated distillation by SFC.

8.1 Introduction

Throughout the history of separation science, the petroleum industry has played a significant role. For instance, in 1897, *Day* [1], while working for the American Geological Survey, developed a method of ascending flow of crude petroleum samples through a column packed with finely pulverized fuller's earth. *Tarasov* [2], in 1926, described the isolation of aromatics from charge stock on silica gel. Shortly after that, the American Petroleum Institute started a project to classify components of petroleum fractions. In more recent times, *Desty,* a British Petroleum chemist, and co-workers built a gas chromatograph based on the original work by *Martin* and *James*. The first International Symposium on Gas Chromatography was held in 1956 and was sponsored by the Institute of Petroleum. During this symposium, Perkin-Elmer introduced the first commercially available GC. A couple of years later, during the Second International GC meeting, *Golay* introduced the concept and applications of open tubular columns. This development revolutionized GC. The renaissance of LC started in the mid-seventies with the advent of high performance column packings and high pressure instrumentation. At present, chromatographic techniques, being LC, GC, TLC or other, are major analytical tools in the petroleum industry.

Although the principles of extraction with supercritical fluids were known since 1879 [3], it was not until 1962 that *Klesper, Corwin,* and *Turner* [4] applied supercritical fluids as mobile phases to the separation of metalloporphyrins. These non-volatile, petroleum derived substances required very high oven temperatures by GC for elution which resulted in sample decomposition. Recent reviews have highlighted the major developments in SFC [5, 6].

For many applications, SFC competes with either HPLC or GC as the preferred analytical method. However, it is now evident that SFC offers real solutions to separation problems such as those dealing with oligomers, polymers and their additives, pesticides, dyes, surfactants, and fossil fuels. This is especially true in cases where certain non-conventional detection modes are required (flame detectors, MS, FTIR) which currently are difficult to realize with HPLC.

In this chapter, applications of SFC in the petroleum industry will be reviewed. We feel that the SFC, like LC and GC, can find a niche in this industry and may establish itself as a routine analytical technique in the near future. The focus in this chapter will be on three sub-areas, *i.e.* hydrocarbon group separations, simulated distillation and the analysis of polynuclear aromatic hydrocarbons (PAH). Although this review deals primarily with SFC, key references in HPLC or GC will also be frequently quoted in order to obtain an accurate picture of current methodology for the analysis of petroleum products. For petroleum applications involving supercritical extraction, the reader is referred to two recently published books [7, 8].

8.2 Hydrocarbon Group Separations

Group type analysis by means of chromatography has been applied to a wide variety of petroleum products including gasolines, naphthas, furnace oils, middle distillates, gas oils, lubricating oils, residues, and total crudes. In the literature, these types of analysis are often abbreviated by the names PONA (paraffins, olefins, naphthenes and aromatics), PIONA (paraffins, isoparaffins, olefins, naphthenes and aromatics), PNA (paraffins, naphthenes and aromatics), PINA (paraffins, isoparaffins, naphthenes and aromatics), or SARA (saturates, aromatics, resins, and asphaltenes). The USBM-API (U. S. Bureau of Mines, American Petroleum Institute) method allows fractionation of petroleum samples into acids, bases, neutral nitrogen compounds, saturates, mono-, di-, and polyaromatic compounds. The analytical separation schemes, mainly based on various forms of liquid chromatography, have been reviewed in 1979 in a book edited by *Altgelt* and *Gouw* [9]. More recently, methods based on GC have been summarized by *Smith* and *Paulsen* [10].

The ASTM D1319 fluorescent indicator analysis (FIA) method has been the standard method for light petroleum fractions (boiling range of 28–315 °C) for more than thirty years. Several problems exist with the FIA method [9]. Recently developed LC methods, designed to improve the scope, speed and accuracy of the FIA method, have gained considerable interest. *Suatoni et al.* [11] used a fluorocarbon (FC-78) mobile phase with silica columns for group separations of gasoline and higher boiling materials (up to 360 °C). A differential refractometer was used for detection. Because response factors vary for individual compounds within a group, the detector needs to be calibrated with standards. The method is currently under investigation by an ASTM study group. *Matsushita et al.* [12] described a system consisting a silver coated silica column to separate olefins from saturates using infrared detection. The problem of calibration remains, however. *Hayes* and *Anderson* [13], recognizing this, developed a system based on dielectric constant detector in which the detector yielded uniformity of response (less than 2.5% relative standard deviation) for each hydrocarbon group independent of the carbon number distribution of the sample. A column consisting of strong cation exchange material in the silver form was used in conjunction with a fluorocarbon mobile phase (freon 123) and a backflush valve to selectivity separate olefins from saturates and aromatics. The method was applied to gasoline, catalytically cracked naphtha, jet fuel, diesel fuel, light pyrolysis fuel oil, and a hydrostabilized fuel oil. In a later paper [14], the same authors described a method to determine naphthenes in gasoline and kerosene having distillation end points of at least 300 °C. Two six-port valves were used with an "unsaturate-selective"-and two "naphthenic-selective" columns. Since the flame ionization detector (FID) gives nearly equal response factors for individual hydrocarbon components, it is an ideal detector to use for group separations. However, FID detection is not suitable for HPLC of lighter petroleum fractions. On the contrary, SFC can solve this detection problem in many cases, as decompressed fluids can be handled quite easily by flame based detectors. Yet the same time, good selectivity can be obtained by using HPLC type stationary phases and suitable fluid systems.

As in the papers mentioned in the above paragraph dealing with HPLC, *Norris* and *Rawdon* [15] used a silver coated column for the paraffin/olefin separation, but in this case SFC was employed to allow FID detection. The method was applied to various petroleum liquids boiling below 350 °C, including gasoline, kerosene, shale oil and creosote oil. Conventional 4.6 mm i. d. packed columns were used while the FID was modified with a split flow arrangement. The method reportedly suffered from poor column-to-column reproducibility due to the procedure followed in making the silver coated silica. Recent results by *Lienne* and co-workers [16] suggest a better control with regard to stationary phase reproducibility.

Schwartz and *Brownlee* [17] described the use of a novel solvent, supercritical SF_6, which has a very low solvent strength, to effect the olefin/paraffin separation of gasoline samples. The critical pressure and temperature of SF_6 are 46.5 °C. and 545 psi, respectively. The aromatic fraction of gasoline was eluted with a simultaneous step-program of pressure and temperature. Microbore (1 mm i. d.) silica columns were used and gasoline samples were injected neat, without dilution, on the column. Although it was observed that SF_6 decomposes in the flame detector, FID detection was made possible by gold plating the collector of the detector. The microbore system was operated without splitting the flow which resulted in excellent peak area and retention time precision. It was found necessary to incorporate a guard column packed with activated (dry) silica to control the water content of the SF_6 which, in turn, had a direct influence on retention times.

Powerful spectrometric techniques such as MS, FTIR, and diode array detection have been used with SFC to confirm identification of certain compounds, or to determine peak homogeneity. Approaches to coupling SFC with FTIR have been reviewed recently by *Jinno* [18] and *Later et al.* [9]. *Hellgeth, Fessehaie,* and *Taylor* [20] used a flow cell-type interface for SFC-FTIR analysis of gasoline and jet fuels. Using the chromatographic conditions of *Schwartz* and *Brownlee* [17], the authors used packed microbore silica columns in conjunction with SF_6 as mobile phase, and compared SF_6 to CO_2 for its utility in FTIR detection. CO_2 has significant pressure dependent bands at *ca.* 1400 – 1100 cm^{-1}. It was found that SF_6 has pressure dependent bands which are comparatively weaker and exhibit a minimal change in infrared transparency (*ca.* 0.02 AU over a 4000 psi pressure increase). Hence, SF_6 proves to be an interesting fluid for use with FTIR detection applied to hydrocarbon group analysis.

The separation of paraffins, olefins, and aromatics in gasoline and middle distillate fuels using supercritical SF_6 was recently also investigated by *Campbell et al.* [21]. The authors proposed two methods to achieve adequate selectivity for the paraffin/olefin groups. One approach uses a differential system, *i.e.* a silica column (1 mm i.d.) was used to isolate and determine the aromatic fraction, while a silver coated strong cation exchange column was used for separation of the paraffins from the olefins and aromatics. Subsequently, the olefins could be determined by the difference of the two runs. In the other approach a column switching scheme was employed which allowed group separation in a single run. The mobile phase was 10% CO_2 in SF_6 and FID detection provided good quantitation and accuracy for various samples. The top trace in **Figure 1** shows a diagram of the proposed column switching method and in the bottom trace of Figure 1 the group separation of a gasoline sample which contained 23.6% olefins is shown. The analysis time is *ca.* 45 min.

Levy and *Ritchey* [22], in a paper dealing with modifiers in SFC, demonstrated the utility of small amounts of formic acid in CO_2 for the group separation of gasoline. Formic acid is one of the few modifiers which can be used in SFC with FID detection, as it gives very little response in this detector. Four 4.6 mm i.d. columns, including a silica-, an amino- and two silver nitrate modified columns were used in series.

Multidimensional techniques, *i.e.* the on-line combination of two or more chromatographic techniques can be very useful to gain further information about the individual components of chemical groups. Compounds can be isolated and identified from complex matrices while detailed "fingerprinting" of petroleum or coal products is feasible. A multi-column approach based on coupled packed GC columns was described by *Boer, Van Arkel,* and *Boersma* [23]. This PNA or "naphtha" analyzer incorporates three columns with different polarities, three traps, four switching valves, and a reactor and is commercially available. Typically, a polar column is used to retard the aromatics. A molecular sieve 13Å column then separates the paraffins and naphthenes according to their carbon number, while a nonpolar column provides further separation of the aromatics and naphthene components. PINA, PIONA, and PONA analyses are feasible in various configurations of the instrument.

A fully automated microbore HPLC-capillary GC technique for the analysis of kerosene and diesel fuels was reported by *Davies et al.* [24]. A retention gap (an uncoated piece of fused silica capillary tubing) was employed to preconcentrated the sample on the capillary GC column. A similar technique was reported by *Munari et al.* for various petroleum materials [25].

The first on-line SFC/capillary GC instrument was recently described by *Levy et al.* [26]. Using this technique, fractions eluting from a packed column can be selectively transferred to a high-resolution capillary GC system. The SFC, which is equipped with UV absorbance and FID detectors in series, can be used for the rapid pre-separation of various hydrocarbon fuels into classes of compounds. Using a

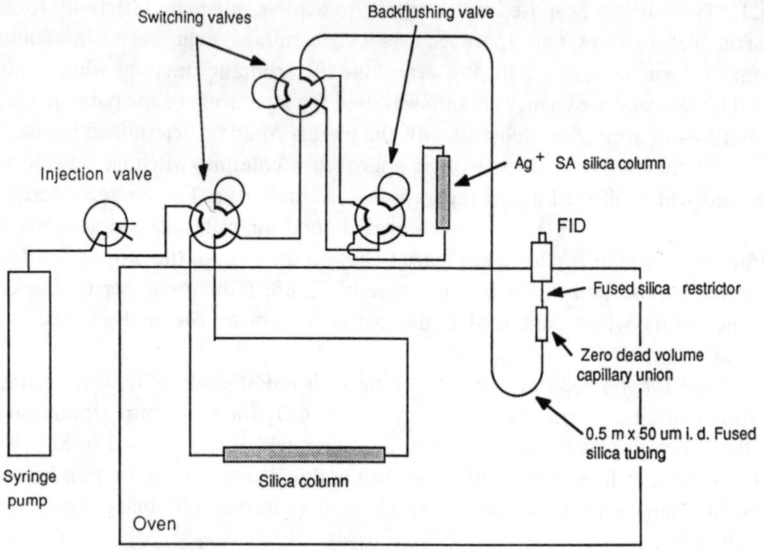

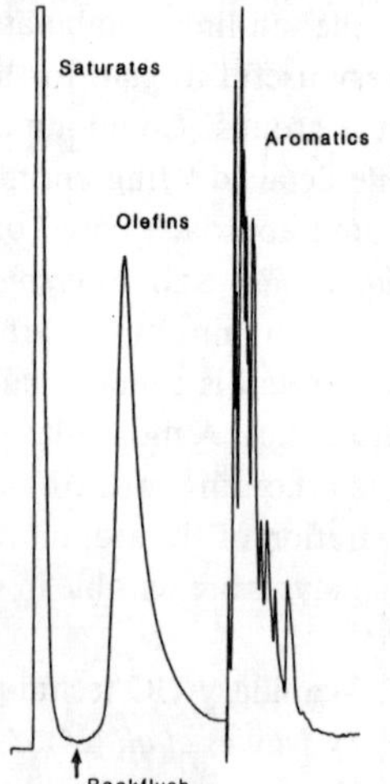

Figure 1
(a) Diagram of a column switching system for group analysis by SFC (b) Group separation of a gasoline sample. Mobile phase: 10 mole % CO_2 in SF_6. Detection FID, 325 °C. Reprinted with permission from ref. [21].

1/32-in. stainless steel transfer line with a high pressure three-port stream selection valve, these hydrocarbon classes can be individually heartcut and sent over to the GC for detailed characterization and quantitation. This allows for the separation of individual components from complex sample matrices which distinctly simplifies subsequent detection and identification. This is shown in **Figure 2** for a high-olefin gasoline sample as an example. The stainless steel transfer line is inserted 40 mm directly into the heated GC injection port (250 °C), which results in maintaining chromatographic integrity as is evident by the sharp peaks of the capillary GC

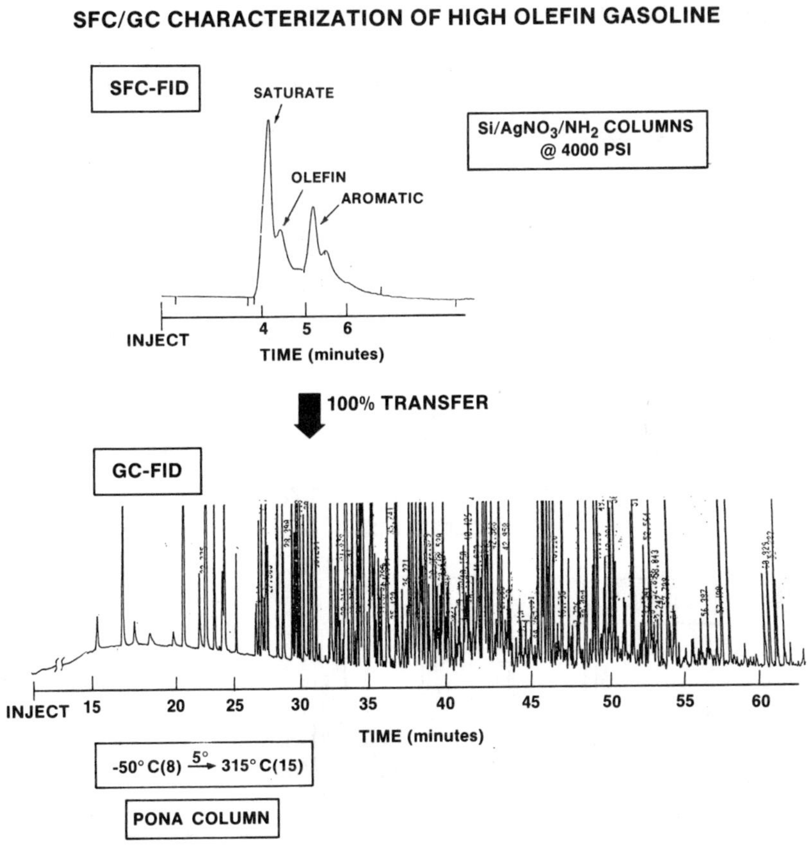

Figure 2
SFC/GC characterization of a high olefin gasoline sample. The SFC system consisted of silica, silverimpregnated silica, and amino columns in series. Mobile phase: CO_2. Column inlet pressure: 4000 psi. Detection: FID, 240 °C. GC: Hewlett-Packard "PONA" capillary column. Temperature program: -50 °C for 8 min, ramp to 315 °C at 5 °/min.

chromatogram. A hot GC injection port is an ideal means for decompression of supercritical fluids because the entire inner volume is heated which minimizes Joule-Thompson cooling effects. The top chromatogram in Figure 2 was obtained using three 25 cm x 4.6 mm i. d., 5-μm packed columns in series. The bottom chromatogram represents complete transfer of the saturate, olefin, and aromatic fractions to a 50m x 0.2 mm i. d. capillary "PONA" column. The GC oven temperature program was initiated at −50 °C to promote focusing of the volatile, early eluting components in gasoline. Since supercritical CO_2 was used, no solvent evaporation was necessary due to the lack of CO_2 response with FID detection. **Figure 3** shows the chromatogram of a heartcut of the olefin fraction from the same high olefin gasoline sample. Compared to the capillary GC chromatogram in Figure 2, the chromatogram in Figure 3 is greatly simplified. This olefin fraction can then be

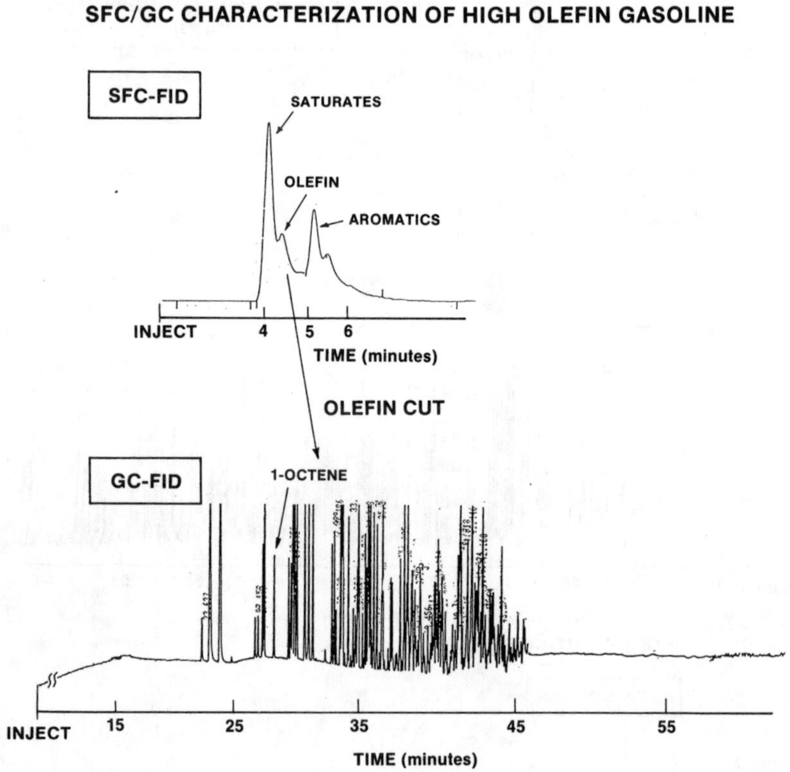

Figure 3
On-line SFC/GC using heart-cutting of the high olefin gasoline sample. Conditions as in Figure 2.

more easily characterized and quantified using FID and mass spectrometric detection. Since the olefin and saturate peaks are not completely resolved on the SFC, some carry-over of saturate components exist in the olefin fraction. However, complementary information on the gasoline can be obtained simultaneously with respect to quantitation of the olefin class on the SFC and selective fingerprinting of the individual olefins on the GC.

High boiling deasphalted residues (boiling point greater than 350 °C) and North Sea oil were examined by *Lundanes* and *Greibrokk* [27]. The SFC system consisted of three 1–2 mm i. d. columns (a cyano, a silica, and a silver-impregnated silica) and two switching valves to allow back-flushing. The silica and the silver-coated silica column produced the separation of the saturates and aromatics, after which the trapped polar compounds were back-flushed from the cyano column. CO_2 was used as mobile phase and FID as detector. Good quantitation was obtained, as was apparent from a comparison of areas obtained by FID with the results obtained by LC and gravimetry. In a later paper [28], the same authors compared CO_2 with N_2O and found that N_2O gave less tailing of the back-flushed aromatic and polar compounds. However, because of a higher background level in the FID, the detection limit with N_2O is three orders of magnitude higher than with CO_2. In addition, the platinum restrictor was clogged easier with N_2O. Therefore, CO_2 was considered the fluid of choice for group separations. **Figure 4** shows a group separation of a high boiling residue into saturates, aromatics, and polar compounds on a system with N_2O (a) and CO_2 (b) as mobile phase, respectively.

The group separation of *n*- and isoparaffins from mineral oils on molecular sieve 5Å columns was investigated by *Ecknig* and *Polster* [29]. Analytical- as well as preparative (combining extraction with chromatographic principles) separations were demonstrated. The selectivity of this separation is based on the trapping of the *n*-paraffins on the molecular sieve while the isoparaffins are unretained. Desorption of *n*-paraffins is typically done under high temperature conditions. In industrial scale separations, the desorption of *n*-paraffins is performed either by means of a displacing agent in the gas phase, by solvents in the liquid phase or by applying a vacuum. In the SFC method of *Ecknig* and *Polster* the isoparaffins are eluted with supercritical CO_2 at relatively low density, after which the *n*-paraffins are eluted with a step pressure program. A key advantage of this procedure is that relatively mild thermal conditions can be employed, minimizing the risk of sample decomposition. The analytical separation was scaled up to fractionate sample sizes of 4 g.

8.3 Analysis of Polynuclear Aromatic Hydrocarbons

Continuous efforts are being made to develop analytical methods for the analysis of PAH in complex matrices. PAH are the largest class of proven chemical car-

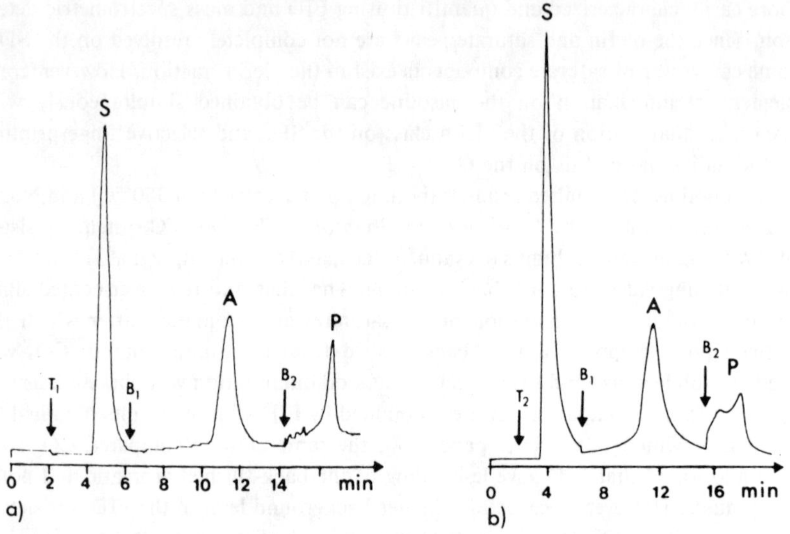

Figure 4
Group separation of a deasphalted residue from North Sea oil into saturates (S), aromatics (A) and polar silica and cyano column system (27). The aromatics left the cyano column at T1 (chromatogram a) and T2 (chromatogram b); backflushing of the aromatics started at B1 and of the polar compounds at B2. Detection: FID, 350 °C. (a) mobile phase: NO_2; (b) mobile phase: CO_2. Reprinted with permission from ref. [28].

cinogens and it is therefore important that they be monitored as environmental pollutants. Due to their fused aromatic ring structures, numerous PAH isomers are possible, especially as their molecular weight increases. Moreover, they are present in such complex matrices as hydrocarbon fuels and waste streams in varying concentrations. Due to environmental concerns, levels of detection for PAH need to be in the low part per million (< 5 ppm) range.

PAH have been analyzed with SFC since the earliest studies of SFC as an analytical technique. PAH ranging from 2 to 6 rings were used as model test compounds by *Sie* and *Rijnders* [30, 31] for theoretical studies. They also applied SFC to the analysis of PAH in coal and polyphenyl tars [32]. *Jentoft* and *Gouw* [33, 34] separated PAH standard mixtures and determined PAH in automobile exhaust by SFC. *Gere et al.* [35, 36] demonstrated rapid SFC analyses (less than 10 min) with highly efficient, packed columns, and CO_2 as mobile phase. The effect of modifiers was also examined by these workers, as well as by several other researchers [22, 37–39]. *Christensen* [40], using CO_2 as mobile phase, described a multidimensional separation with valve switching for coal tar samples. Amino- and octadecylsilane columns were employed to separate fractions containing 16 and 18 aromatic carbons. *Campbell* and *Lee* [41] used an aminosilane phase bonded on 30–70 μm silica particles and CO_2 as mobile phase to fractionate various petroleum and coal derived mixtures. The relatively large particle size of the packing material minimizes the pressure drop across the column. The fractions were collected in pressurized vessels for subsequent analysis by capillary GC.

Capillary SFC has also been used by Lee and co-workers [42, 43] to provide high resolution analyses of PAH in complex samples. Supercritical pentane was used as mobile phase at 210 °C in conjunction with a 50% phenyl polysiloxane capillary column and fluorescence detection. In a recent publication, *Kuei et al.* [44] described a SFC system in which supercritical ammonia was used as mobile phase. It was found only an octyl modified polysiloxane coated capillary column was stable under these conditions. Examples of the analysis of PAH in a carbon black extract were shown. *West* and *Lee* [45] used capillary SFC with thermionic detection for the analysis of thermally labile hydroxynitropyrenes. Mass spectrometry has been a useful tool in the identification of PAH in complex samples. The interfacing of SFC with MS has been detailed by researchers at Batelle Northwest [46]. *Wright* and *Smith* [47] used capillary SFC for the analysis of nitrogen-containing PAH fractions and hydroxylated PAH fractions from middle distillate fuels. In another paper [48], the analysis of PAH in selected marine diesel fuel samples by capillary SFC-MS was described. **Figure 5** shows an example of a total ion chromatogram generated by isobutane chemical ionization. A key factor in the analysis of PAH in hydrocarbon streams is the complexity of the matrix. GC is restricted by the low volatility of the large molecular weight PAH while HPLC lacks the resolving power. **Figure 6** shows an example of packed column SFC being utilized to separate four and five fused ring PAH from a heavy hydrocarbon stream in less than 9 min. This was accomplished with a 25 cm x 4.6 mm i. d., 5-μm C-18 column, UV detection, and a CO_2 mobile phase which was modified with propylene carbonate [49]. The use of modifiers such as propylene carbonate enhances supercritical mobile phase selectivity by varying PAH retentivity with respect to the hydrocarbon matrix. Additional selectivity can be achieved by varying not only the modifier concentration, but also the modifier identity [49]. This is shown graphically in **Figure 7** with a plot of PAH capacity factor (from naphthalene up to coronene) *versus* modifier identity. The capacity factors were determined by using a 25 cm x 4.6 mm i. d. diol (10 μm) column with all of the modifiers being at an equivalent concentration of six mole percent [49]. Carbon dioxide with no modifier added is shown as a reference. As can be seen, propylene carbonate, dimethylacetamide, and dimethyl sulfoxide provide for the fastest elution of PAH. Using tetrahydrofuran or methanol as modifiers, PAH exhibit higher capacity factors and are significantly more retained. To achieve intermediate PAH resolution, 1-hexanol or acetonitrile can be used to modify the carbon dioxide.

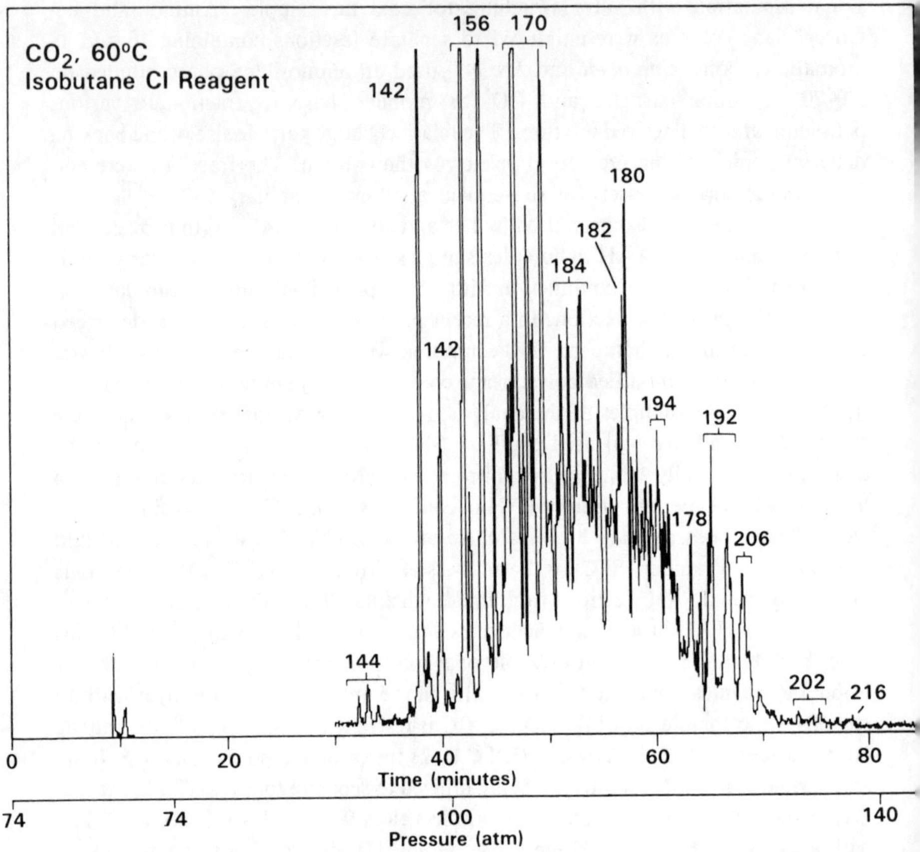

Figure 5
Capillary SFC-MS total ion chromatogram obtained from the PAH fraction of a marine diesel fuel. Isobutane was used as the chemical ionization reagent gas. Peak labels refer to the molecular weights of the compounds. Reprinted with permission from ref. [48].

With various hydrocarbon matrices, the use of modifiers coupled with UV detection presents a distinct advantage in terms of flexibility in method development for different PAH separations.

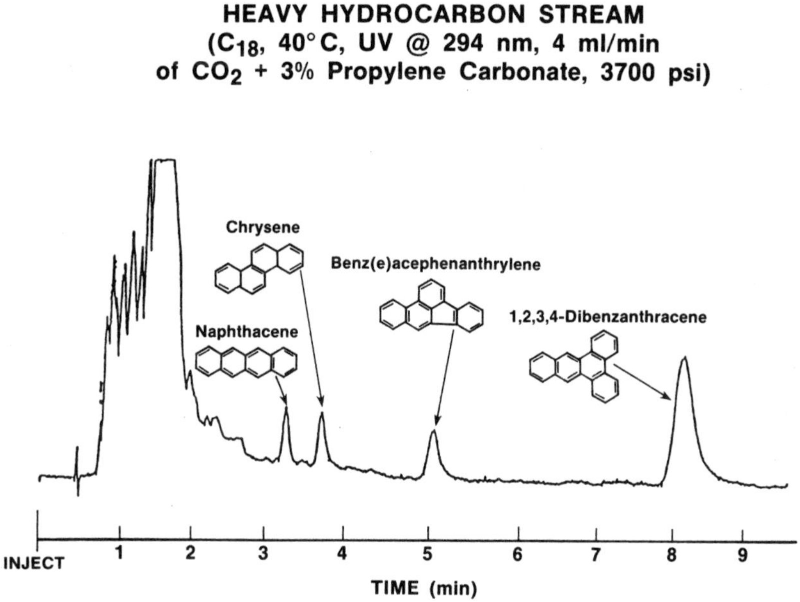

Figure 6
Separation of 4- and 5-ring PAH from a heavy hydrocarbon stream. Mobile phase: CO_2 + 3% propylene carbonate. Column: 25 x 0.46 cm, C_{18}, 5 μm particle size. Oven temperature: 40 °C. Column inlet pressure: 3700 psi. Flow rate: 4 ml/min. Detection: UV, 294 nm.

8.4 Simulated Distillation (SIMDIS)

Simulated distillation by GC is applied in the petrochemical industry to obtain true boiling point distributions of distillates and crude oils. In addition, GC is often used as a "fingerprinting" technique for various crude oils, products, and intermediate streams. Pattern recognition techniques based on cumulative area *vs.* time data are used routinely by refineries for identification purposes [50]. Two standardized methods, ASTM D-2887 and ASTM D-3710, are available for the boiling point determination of petroleum fractions and gasoline, respectively. SIMDIS-GC has been reviewed by *Butler* [9]. The applicability to coal-derived liquids has also been demonstrated [51]. The ASTM D-2887 method utilizes non-polar, packed GC columns in conjunction with FID detection. The upper limit of the boiling range covered by this method is *ca.* 1000 °F atmospheric equivalent boiling point (AEBP). Recent efforts in which high temperature GC were used have focused on extending the scope of the ASTM D-2887 methods for higher boiling petroleum materials.

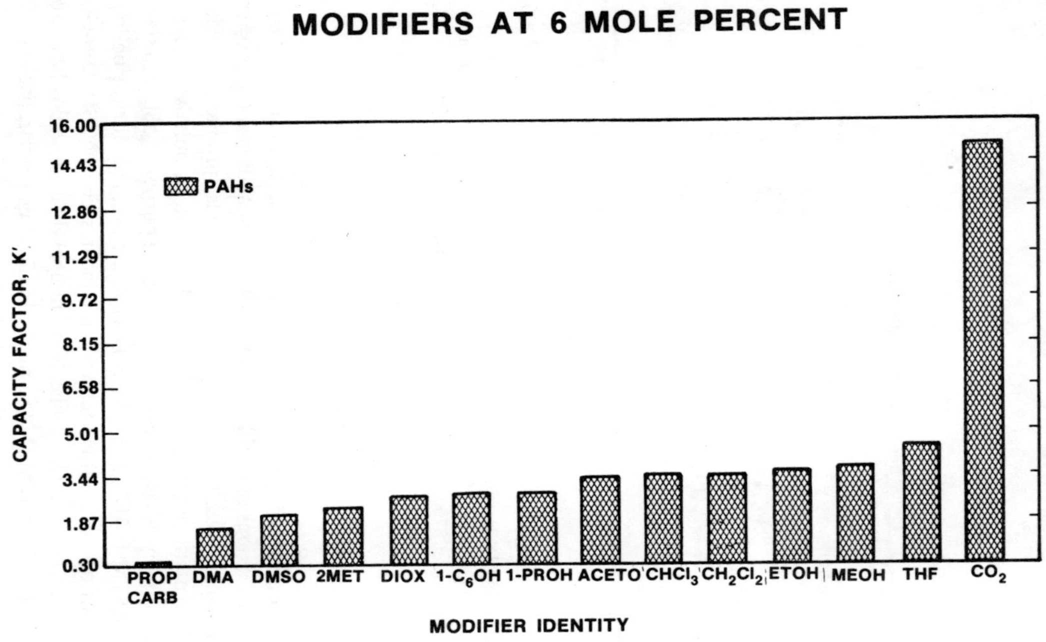

Figure 7
Plot of capacity factor (from naphthalene to coronene) *vs.* modifier identity. CO_2 is shown as a reference. Modifier concentration: 6 mole %. Column: 20 x 0.46 cm, Diol modified silica, 10 μm particle size. Oven temperature: 40 °C. Flow rate: 4 ml/min. Column inlet pressure: 3600 psi. Modifiers are abbreviated as follows: PROP CARB: propylene carbonate. DMA: dimethylacetamide. DMSO: dimethyl sulfoxide. 2 MET: 2-methoxyethanol. DIOX: dioxane. 1-C_6 OH: 1-hexanol. 1-PROP: 1-propanol. ACETO: acetonitrile. $CHCl_3$: chloroform. CH_2Cl_2: methylene chloride. ETOH: ethanol. MEOH: methanol. THF: tetrahydrofuran.

Trestianu et al. [52] demonstrated that glass capillary columns combined with cold, on-column injection can be applied to petroleum fraction ranging from 300 to 1470 °F AEBP. In this approach, column temperatures of up to 430 °C are required for elution of high boiling samples.

In recent publications, *Schwartz et al.* [53, 54] proposed the use of SFC for SIMDIS of high boiling petroleum fractions. The key advantage of SFC compared to GC are the mild elution temperatures required for sample elution. Typically, elution in SFC-SIMDIS is achieved by means of pressure- or density programming and at temperatures of 100–150 °C. Boiling point distributions and reconstructed area chromatograms of petroleum materials are shown in **Figure 8**. The upper trace represents an atmospheric residuum from Altamont crude oil, obtained by removing low-boiling material (less than 650 °F) from the crude by distillation at atmospheric

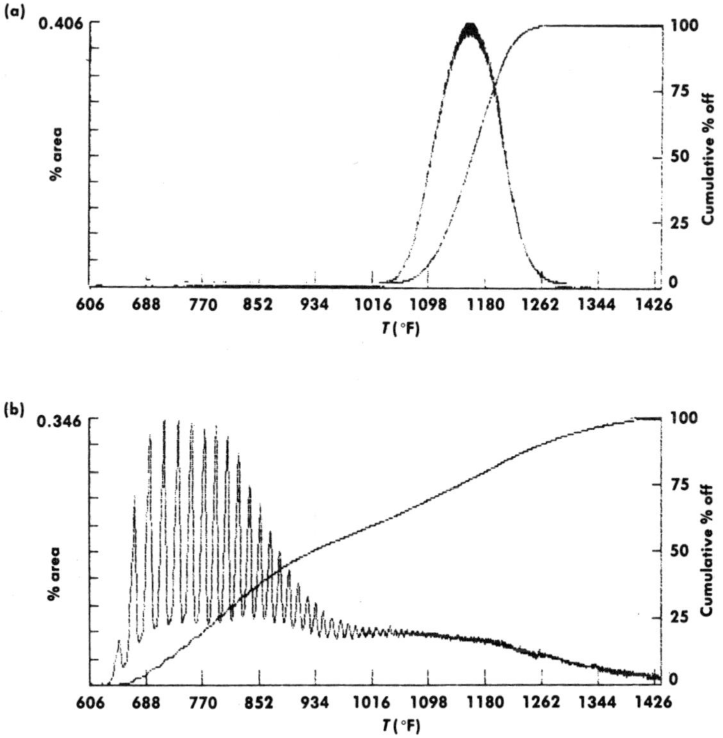

Figure 8
Boiling point distribution and reconstructed area chromatograms of petroleum samples as generated by capillary SFC-SIMDIS. (a) DISTACT fraction of Altamont crude oil (b) atmospheric residuum of Altamont crude oil. Reprinted with permission from ref. [53].

pressure. The lower trace depicts a short-path vacuum distillate (DISTACT) cut from the feed stock. In these examples 100% sample recovery was obtained for the DISTACT fraction; with the atmospheric residuum, only 3% did not elute from the column. The authors also investigated the utility of vacuum thermogravimetric analysis (VTGA) for simulated distillation. It was found that decomposition of samples occured when temperatures exceeding 370 °C were used. In the case of an atmospheric residue from Arabian heavy crude oil, bimodal thermogravimetric profiles were obtained. Presumably, the bimodal distribution is indicative of sample decomposition, which was confirmed by other TGA-GCMS experiments. These findings raise concern about the validity of high temperature GC to SIMDIS analysis, as similar bimodal profiles have been published in recent reports in which high temperature GC-SIMDIS was used [55].

Packed as well as capillary columns can be applied to SFC-SIMDIS [56]. It has been pointed out in recent papers [57, 58] that capillary columns are most useful when the separation takes place under mobile phase conditions of low density. Packed columns on the other hand, are more useful when high density conditions prevail. **Figure 9** shows a chromatogram of Polyethylene-740, a calibration standard used for SIMDIS analysis. In this example, a 10 x 0.1 cm microbore packed column was used at 150 °C. The column material consisted of 15% phenyl, 85% methylpolysilocane

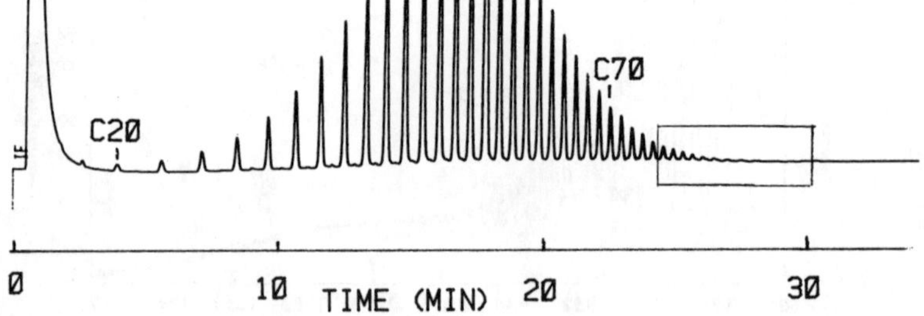

Figure 9
Packed column SFC of a calibration standard, polyethylene-740. Mobile phase: CO_2. Column: 10 x 0.1 cm, polysiloxane coated silica, 7 μm particle size. Oven temperature: 150 °C. Pressure gradient: 2000 psi for 4 min, linear ramp to 5500 psi in 21 min. Injection: 0.5 μl in CS_2. Detection: FID, 350 °C.

modified silica (7 μm particles). Peaks corresponding to alkane carbon numbers of up to C-108 can be discerned while *ca.* 45 peaks eluted within 30 min. A recent paper [59] gives further details on packed column SFC applied to SIMDIS.

References

[1] *D. T. Day,* Proc. Am. Phil. Soc. **36** (1897) 889.
[2] *B. C. Tarasow,* Azerb. Neft. Khoz. **T1** (10) (1926) 47.
[3] *J. B. Hannay,* and *J. Hogarth,* Proc. R. Soc. London **29** (1879) 324–326.
[4] *E. Klesper, A. H. Corwin,* and *D. A. Turner,* J. Org. Chem. **27** (1962) 700–701.
[5] *D. R. Gere,* Science **222** (1983) 253–259.
[6] *C. M. White* and *R. K. Houck,* HRC & CC **9** (1986) 4–17.
[7] "Supercritical Fluid Technology", Process Technology Proceedings 3, *J. M. L. Penninger, M. Radosz, M. A. McHugh,* eds., Elsevier, Amsterdam (1985).
[8] "Supercritical Fluids", Chemical and Engineering Principles and Applications, *T. G. Squires* and *M. E. Paulaitus,* eds., American Chemical Society, Washington (1987).
[9] "Chromatography in Petroleum Analysis", *K. H. Altgelt* and *T. H. Gouw,* eds., Marcel Dekker, New York (1979).
[10] *E. F. Smith* and *K. E. Paulsen* in "Modern Practice of Gas Chromatography", *R. L. Grob,* ed., Wiley, New York (1985) pp. 631–721.
[11] *J. C. Suatoni, H. R. Garber,* and *B. E. Davis,* J. Chromatogr. Sci. **13** (1975) 367–371.
[12] *S. Matsushita, Y. Tada,* and *T. Ikushige,* J. Chromatogr. 208 (1981) 429–432.
[13] *P. C. Hayes, Jr.* and *S. D. Anderson,* Anal. Chem. **58** (1986) 2384–2388.
[14] *P. C. Hayes, Jr.* and *S. D. Anderson,* J. Chromatogr. **387** (1987) 333–346.
[15] *T. A. Norris,* and *M. G. Rawdon,* Anal. Chem. **56** (1984) 1767–1769.
[16] *M. Lienne, P. Gareil, R. Rosset, J. F. Husson, M. Emmelin,* and *B. Neff,* J. Chromatogr. **395** (1987) 255–270.
[17] *H. E. Schwartz,* and *R. G. Brownlee,* J. Chromatogr. **353** (1986) 77–93.
[18] *K. Jinno,* Chromatographia **23** (1987) 55–62.
[19] *D. W. Later, D. J. Bornhop, E. D. Lee, J. D. Henion,* and *R. C. Wieboldt,* LC.GC **5** (1987) 805.
[20] *J. Hellgeth, M. Fessehaie,* and *L. T. Taylor,* Pittsburgh Conference on Analytical Chemistry and Spectroscopy, Atlantic City, N. J., 1987, paper no. 1047.
[21] *R. M. Campbell, N. M. Djordjevic, K. E. Markides,* and *M. L. Lee,* Anal. Chem., in press.
[22] *J. M. Levy* and *W. M. Ritchey,* J. Chromatogr. Sci. 24 (1986) 242–248.
[23] *H. Boer, P. Van Arkel,* and *W. J. Boersma,* Chromatographia **13** (1980) 500.
[24] *I. L. Davies, M. W. Raynor, P. T. Williams, G. E. Andrews,* and *K. D. Bartle,* Anal. Chem. **59** (1987) 2579.
[25] *F. Munari, A. Trisciani, G. Mapelli, S. Trestianu, K. Grob, Jr.,* and *S. M. Colin,* HRC & CC **8** (1985) 2579.
[26] *J. M. Levy, J. P. Guzowski,* and *W. E. Huhak,* HRC & CC **10** (1987) 337–341.
[27] *E. Lundanes* and *T. Greibrokk,* J. Chromatogr. **349** (1985) 439.
[28] *E. Lundanes, B. Iversen* and *T. Greibrokk,* J. Chromatogr. **366** (1986) 391–395.
[29] *W. Ecknig* and *H. J. Polster,* Sep. Sci. **21** (1986) 139–156.
[30] *S. T. Sie* and *G. W. A. Rijnders,* Sep. Sci. **2** (1967) 729–753.
[31] *S. T. Sie* and *G. W. A. Rijnders,* Anal. Chim. Acta **38** (1967) 31–44.

[32] *S. T. Sie* and *G. W. A. Rijnders,* Sep. Sci. **2** (1967) 755–777.
[33] *R. E. Jentoft* and *T. H. Gouw,* J. Chromatogr. Sci. **8** (1970) 138–142.
[34] *R. E. Jentoft* and *T. H. Gouw,* Anal. Chem. **48** (1976) 2195–2200.
[35] *D. R. Gere, R. Board,* and *D. McManigill,* Anal. Chem. **54** (1982) 736–740.
[36] *D. R. Gere,* Application Note AN 800-1, Hewlett-Packard, 1983.
[37] *M. Novotny, W. Bertsch,* and *A. Zlatkis,* J. Chromatogr. **61** (1971) 17–28.
[38] *J. M. Levy* and *W. M. Ritchey,* HRC & CC **8** (1985) 503–509.
[39] *B. W. Wright* and *R. D. Smith,* J. Chromatogr. **355** (1986) 367–373.
[40] *R. G. Christensen,* HRC & CC **8** (1985) 824–828.
[41] *R. M. Campbell* and *M. L. Lee,* Anal. Chem. **58** (1986) 2247–2251.
[42] *J. C. Fjeldsted, B. E. Richter, W. P. Jackson,* and *M. L. Lee,* J. Chromatogr. 279 (1983) 423–430.
[43] *W. P. Jackson, R. C. Kong,* and *M. L. Lee,* Polynuclear Aromatic Hydrocarbons: Mechanisms, Methods and Metabolism Battelle Press, Columbus, Ohio (1985) pp. 609–625.
[44] *J. C. Kuei, K. E. Markides,* and *M. L. Lee,* HRC & CC **10** (1987) 257–262.
[45] *W. R. West* and M. L. Lee, HRC & CC **9** (1986) 161–167.
[46] *R. D. Smith* and *H. R. Udseth,* Anal. Chem. **59** (1987) 13–22.
[47] *B. W. Wright* and *R. D. Smith*, Chromatographia **18** (1984) 542–545.
[48] *B. W. Wright, H. R. Udseth, R. D. Smith,* and *R. N. Hazlett,* J. Chromatogr. **314** (1984) 253–262.
[49] *J. M. Levy* and *J. P. Guzowski,* submitted for publication.
[50] *D. J. Abott,* HRC & CC **9** (1986) 598.
[51] *R. B. Panell* and *A. Sood,* J. Chromatogr. Sci. **20** (1982) 433.
[52] *S. Trestianu, G. Zilioli, A. Saroni, C. Saravalle, F. Munari, M. Galli, G. Gaspar, J. M. Colin,* and *J. L. Jovelin,* HRC & CC **8** (1985) 771.
[53] *H. E. Schwartz, J. W. Higgins,* and *R. G. Brownlee,* LC.GC **4** (1986) 639–643.
[54] *H. E. Schwartz, R. G. Brownlee, M. M. Boduszynski,* and *F. Su,* Anal. Chem. 59 (1987) 1393–1401.
[55] *A. D. Bashall,* Am. Lab. **19** (1987) 56.
[56] *H. E. Schwartz, P. J. Barthel, S. E. Moring,* and *H. H. Lauer,* 11th Intl. Symposium on Column Liquid Chromatography, Amsterdam, 1987, paper no. Tu-L-12.
[57] *H. E. Schwartz, P. J. Barthel, S. E. Moring,* and *H. H. Lauer,* LC.GC **5** (1987) 490–497.
[58] *P. J. Schoenmakers* and *L. G. M. Uunk,* Europ. Chromatogr. News **1** (1987) 14.
[59] *H. E. Schwartz,* J. Chromatogr. Sci. **26** (1988) 275–279.

9 Analytical and Applied Separations with Supercritical Fluids and Alumina

Robinson P. Khosah
Analytical Chemistry Division, Alcoa Laboratories, Alcoa Center, PA 15069, USA

Dr. Robinson P. Khosah is currently a Staff Scientist at Alcoa Laboratories. A citizen of Zimbabwe, he received his BSc in chemistry and botany from the then University of Rhodesia in 1976. After working briefly as an analytical chemist at Tinto Industries' iron and steel foundry, he came to the USA for graduate studies. Under the direction of *Prof. Peter Uden* at the University of Massachusetts at Amherst, he obtained an MS degree (1981) and a PhD (1984) degree in analytical chemistry. His areas of concentration were pyrolysis/gas chromatography/mass spectroscopic studies of fossil fuel materials and coupling of selective or element specific detectors in gas or liquid chromatography.

Since joining Alcoa in 1984, he has served in various capacities in GC and/or MS, as applied to Environmental Protection Agency and Food and Drug Administration regulated analyses and industrial applications of pyrolysis/GC/MS. He has also worked toward the fundamental understanding of the nature and analysis of hydrogen gas and inclusions present in aluminum and its alloys. Since 1985, he has spearheaded and directed research on supercritical fluid chromatography (SFC) and other applications of supercritical fluid technology. Notable success areas include the application of alumina and polymer packed columns for SFC analysis, the modeling of pilot scale adsorption/desorption supercritical fluid separations and purification of polymers using supercritical fluids.

Dr. Khosah is a member of the American Chemical Society, the Society for Analytical Chemists of Pittsburgh and an affiliate member of IUPAC.

Summary:

The recent renewed interest in supercritical fluid technology has spurred a considerable amount of research into both the fundamental and practical aspects of these fluids and their applications in separations. This chapter will explore the practical applications of supercritical fluid chromatography (SFC) using both fused silica capillary and alumina packed columns.

Various model compounds were used to evaluate the column and instrument performance. Conditions were established to elute and resolve oligomeric series of polyethylene glycols, polypropylene glycols, and Triton X-100, and to distinguish between series of different average molecular weights. The effects of changing: i) the pressure and temperature of the mobile phase; ii) the column type, length and diameter; and iii) the lengths and diameters of splitter and restrictor interfaces, on the elution and resolution of solute moieties were also determined. Once the operating conditions had been established, "real-world" sample analyses were performed.

The applications included analyses of poly (α-olefins), lubricant materials and fish oils. Separations were deemed feasible with both open tubular and packed columns. Columns packed with neutral alumina were especially successful with hydrocarbon or non-polar oligomer separations. Work was also performed on polymer purification with a supercritical fluid extraction/supercritical fluid chromatography (SFE/SFC) unit.

Concentrate:

Supercritical fluid chromatography was shown to be a good technique for the analysis of high molecular weight or polar compounds found in lubricants and fish oils. Alumina (neutral) packed colums were also shown to provide good separations of non-polar hydrocarbon and polymer materials.

Pilot scale separations or polymer purification can also be modeled using a bench scale SFE/SFC unit.

9.1 Introduction

Supercritical fluids (SCF) technology is a term that embodies the various separation processes that utilize supercritical fluids and the field of reactions that are performed under supercritical fluid conditions. The separation processes have, unfortunately, been classified under various acronyms and this has tended to split the efforts of researchers trying to achieve the same goal.

The extraction of components from liquid or solid matrices by SCF is termed supercritical fluid extraction (SFE or SCFE). The separation of compounds at the analytical level by chromatography is termed SFC. Those researchers who combine both techniques also have their acronym, SFE/SFC. Important information can be gathered from SFC or SFE/SFC that can be used for SFE and *vice versa.*

The term supercritical fluid is used to describe a highly compressed gas that is near or above its critical temperature and critical pressure points. The theory of supercritical fluids and supercritical fluid chromatography has been well covered in the previous chapters. A synopsis of the properties of supercritical carbon dioxide is, however, being presented here since this was the principal mobile phase used.

Figure 1 is a phase diagram of carbon dioxide (CO_2) that depicts its triple point, critical temperature (31.1 °C), and critical pressure (72.8 atmospheres) It has been determined that at the critical temperature the vapor and liquid phases of CO_2 have the same densities. Above its critical temperature, irrespective of the pressure, CO_2 cannot be liquefied and it exists as a supercritical fluid. The physical proper-

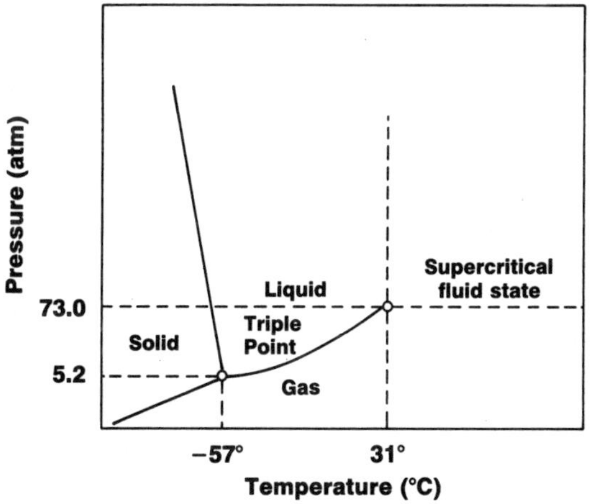

Figure 1
Phase diagram of carbon dioxide.

ties of a supercritical fluid in general are intermediate between those of the gas and liquid states. This imparts to the fluid unique solvent powers similar to those of a liquid, along with the diffusivity of a gas. This and the fact that the viscosities are generally lower than those of liquids and most of the physical and chemical properties are a function of the density, make supercritical fluids versatile and selective separation media [1].

There are other gases and liquids besides carbon dioxide that can attain supercritical fluid conditions at moderate temperature and pressure conditions. Carbon dioxide is often the fluid of choice given its low critical parameters and its existence as a colorless, odorless, and non-toxic gas that can be obtained at high purities. Due to its non-polarity, its solvent capabilities are enhanced by the addition of modifiers, especially where polar compound separations are involved. Low molecular weight alcohols such as methanol are used as modifiers for carbon dioxide. Thus, in selecting a supercritical fluid for a specific application, factors such as solute polarity, solubility in the fluid, chemical inertness of fluid, and the temperature and pressure stability of the fluid all have to be taken into account.

The first reported use of a supercritical fluid to dissolve solid materials was by *Hannay* and *Hogarth* at a meeting of the Royal Society of London in 1879 [2]. They demonstrated that inorganic salts such as cobalt chloride or potassium bromide could be made to dissolve in or precipitate out of ethanol by altering the pressure at temperatures above 234 °C, which is the critical temperature of ethanol. Seventy-five years later, in 1954, *A. W. Francis* [3] reported the solubilities of 261 compounds in near-critical liquid carbon dioxide (25 °C and 950 psi).

The first documented proposal of the use of supercritical fluids as chromatographic mobile phases was in 1958 in a notarized document by *James E. Lovelock* at Yale University. In that document, mention is made of the potential for using critical state chromatography for the analysis of ionic compounds. The first experimental results were published in 1962 by *Klesper et al.* [4]. These authors documented the separation of nickel etioporphyrin isomers using supercritical dichlorodifluoromethane and monochlorodifluoremethane. Subsequent findings were reported by *Klesper* and *Hartmann* [5] using SFC with packed columns to separate polystyrene oligomers.

Sie and *Rijnders* et al. [6–11] also showed SFC applications with various mobile phases and packed columns for the analysis of polyaromatics, dyes, resins, and antioxidants. The work of *Giddings et al.* [12, 13] exemplified further the applications of SFC for the analysis of polypeptides, amino acids, carotenoids, and various polymers such as methylsilicones and polyethylene glycols. A decade went by without many SFC publications. The resurgence of interest in SFC in the early 1980s was mostly spurred by the work of *Novotny* and *Lee*, which resulted in a patent application on open-tubular supercritical fluid chromatography in 1982. That patent was awarded in 1984. Several SFC applications were also demonstrated by *Gere* [14] using packed columns and a converted Hewlett Packard high performance liquid chromatograph (HPLC).

The first commercially available capillary SFC unit with capabilities for packed column SFC, however, was produced by Suprex Corporation in 1985. The emphasis by Suprex on demonstrating applications [15, 18] and the interest generated was followed by to the production of SFC instruments by other companies, *e. g.*, Lee Scientific, Inc., Brownlee Labs and Carlo Erba, and a significant increase in SFC related publications.

The interest in supercritical fluid extraction has also increased dramatically. This topic is covered thoroughly in a book by *McHugh* and *Krukonis* [19].

Supercritical fluid chromatrography was introduced to Alcoa Labs as an alternative technique for the analysis of samples, such as some lubricant materials or additives, that are difficult to analyze by regular chromatographic methods. Successful analyses were obtained with some proprietary lubricant formulations and polymer materials, *e. g.*, poly (α-olefins). These analyses were obtained with capillary SFC using gas chromatography (GC) columns. Another analytical finding was that some components of fish oils could be resolved by SFC.

Realizing that the composition of some of these oils is generally non-polar hydrocarbons mixed with polar acids and esters, it was decided to try the same SFC separation, but this time using columns packed with regular alumina. As expected, the only components eluted from the columns were non-polar hydrocarbons, with the rest of the matrix remaining on the column. The benefits were apparent: (a) alumina could be used in packed columns for the analytical SFC separation of hydrocarbon based materials, (b) alumina with supercritical fluids could be used for selective or preferential adsorption and desorption of compounds, especially non-polar moieties in a polar matrix.

Applications are still being developed with functionalized alumina. SFC is also being used to model pilot scale separations by performing adsorption/desorption experiments on an SFE/SFC unit. The data obtained can then be tested on a pilot scale unit to determine the commercial feasibility of various processes.

9.2 Experimental Procedure

9.2.1 Apparatus

The equipment used was obtained from Suprex Corporation, Pittsburgh, PA. The analytical SFC unit is a Model 200A equipped with two flame ionization detectors and capable of performing both capillary and packed column SFC. The SFE/SFC unit was designed at Alcoa and custom-built by Suprex, and it has capabilities to perform material extraction or desorption from media with supercritical fluids.

Simultaneous analysis of the eluted products is achievable in addition to real time monitoring of the total extractives. With minimum rearrangement, the unit can be converted into two SFC units to perform analytical work or testing of columns.

9.2.2 Analytical Work

9.2.2.1 Capillary Column Techniques

On obtaining the first analytical unit, the problems generally observed with split injection capillary SFC were encountered. These problems were associated with the splitter and restrictor interfaces. Studies were performed to determine: (i) the effects of restrictor length, tube diameter and tip diameter: (ii) splitter length and diameter; (iii) column positioning; (iv) column type, diameter and length; and (v) operating parameters with various samples.

The samples analyzed included polypropylene glycols (PPG), polyethylene glycols (PEG), poly(α-olefins) (PAOs), lubricant materials, Triton X-100, and fish oils.

9.2.2.2 Packed Column Techniques

The studies performed with packed column SFC were similar to those for capillary SFC. The differences were due to: (i) no splitter interface was required, and (ii) the arrangement for the restrictor interface was different. The samples analyzed were also different, as will be explained later.

9.2.3 Adsorption/Desorption and Purification Studies

The experiments performed were designed to provide desorption data on materials previously adsorbed on alumina or for the removal of contaminants from materials. The samples studied were polyester materials and results obtained, on the selective removal of their contaminants with supercritical carbon dioxide, will be reported.

9.3 Results and Discussion: Analytical Work

9.3.1 Capillary Columns

The data reported for this part reflect some of the work performed to evaluate the equipment. They also provide understanding of the fundamentals of SFC.

9.3.1.1 Sample Introduction

The problems associated with the splitter and column positioning can be related to the difficult nature of sample introduction encountered with capillary SFC. The compounds to be analyzed generally are either of high molecular weight or are polar. Even though they may be soluble in a particular solvent, those compounds may not be soluble in liquid carbon dioxide, which is the state of the mobile phase in most injection ports of SFC units. The supercritical fluid conditions are, in general, established in the oven where the separation column is placed. Consequently, the sample precipitates out in the injector or splitter components. Several remedies were tried, some with success.

9.3.1.1.1

Heating the injector head was found to aid in sample introduction. The only problem was that higher boiling solvents became necessary where volatile solvents such as methylene chloride would have been sufficient. The latter solvents tended to bubble, even with heat from the oven, which nullified the benefits realized from heating the injector.

9.3.1.1.2

Less frequent plugging of the splitter was observed when the splitter was placed within the heated oven. Since the splitter regulates the amount of sample introduced onto the column, the diameter and length of the splitter determine the split ratio. A 60 cm x 10 μm i. d. untapered fused silica tubing provided optimum splitter functions and minimal clogging.

9.3.1.1.3

To obtain the optimum column position, the column was pulled back a distance, not to exceed 1/8 ″, from the injector exit. An indication of an improperly positioned column is a jagged trailing edge on a solvent peak. This can arise from a column positioned too close to the exit port, resulting in column tip damage by the rotor. At distances greater than 1/8 ″, it is also suspected that in addition to band broadening, the presence of "eddy currents" in the splitter assembly results in uneven introduction of sample onto the column. This effect could also result in memory effects where previous material deposited on the walls of the splitter assembly are redissolved by the new solvent and introduced onto the column.

The factors described above are not easily defined, and their significance is best appreciated by experience.

9.3.1.1.4

Columns – Several capillary columns of various types (functionality), length, diameter and film thickness were tested for SFC compatibility. In general, DB-225, DB-5, and DB-17 columns gave good separations for most compounds ranging from polar to non-polar. DB-WAX columns were determined to be the best columns for the SFC of polar compounds using carbon dioxide as the mobile phase.

9.3.1.2 Film Thickness and Diameter

The optimum film thickness for most of our separations was 0.2 μm. Greater film thicknesses (> 0.25 μm), especially with DB-WAX, resulted in very long retention times and broad peak shapes. As with GC, the smaller the internal diameter of the column, the higher the efficiency of separations. The splitter problems enumerated above are, however, more noticeable with smaller i.d. columns. The aim should, therefore, be to use the largest diameter column possible, which may eliminate the use of an injection splitter, and improve resolution by increasing the length of the column used. Megabore (530 μm i.d.) columns should not be used because of poor wall strengths; also, they shatter to powder, even at low pressures.

9.3.1.3 Length

The optimum column length was found to be around 10 meters. The gain in resolution obtained with longer columns is partially offset by the increase in retention times. Since most SFC separations of interest are for high molecular weight compounds, such analyses can indeed be achieved with good resolution and short retention times using very short columns (2–5 meters).

9.3.1.4 Operating Parameters

The optimum parameters required for separations will be detailed for individual sample analyses. However, the general observation is that separations are based on the density of the mobile phase. This parameter is altered by adjusting the pressure or temperature of the supercritical mobile phase. There are times when it is necessary to perform negative temperature programming in order achieve higher densities, at a given pressure, and achieve better separations. There are also instances when better resolution is achieved by ramping the pressure up to a preset value and increasing the temperature. This is applicable to sample with compounds of a large range in molecular weight and/or volatility. Although SFC operating parameters are sample specific, initial separation conditions can be established from knowledge of the molecular weight range or the volatility range of sample components. It was also observed that linear pressure ramping worked well with mixture of oligomers, while multiple pressure ramps were needed for more complex mixtures.

9.3.1.5 Sample Analyses

A number of samples of various functionalities and compositions were analyzed to determine compatibility with SFC analysis and column types. PPG, PEG, and Triton X-100 samples ultimately were used as standards to evaluate column and instrument performance. Other samples of interest to Alcoa were evaluated, once optimum operating conditions had been determined.

9.3.1.5.1 Polypropylene Glycol (PPG-425)

Many studies were performed with polypropylene glycol oligomers (av. M. W. = 425). The oligomers were shown to separate well on different columns. **Figures 2a** and **2b** show the effect of temperature and/or density on SFC analysis. The only difference between the two chromatograms is that Figure 2a was obtained at 100 °C and Figure 2b was obtained at 150 °C. The first peak in each chromatogram is the

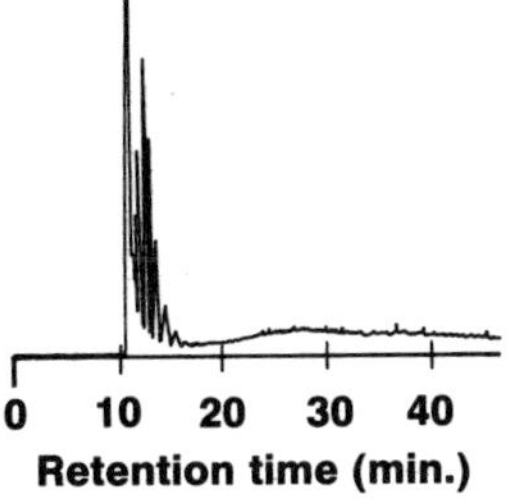

Figure 2a
PPG 425 in methanol; 100 °C; 200 atm/10 min; 200–350 atm/15 min; 350 atm – hold; 10 meter DB-17 column; Atn. = 4; Chart speed = 0.2 cm/min.

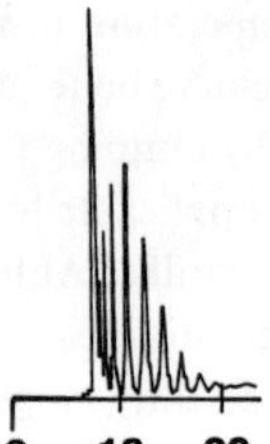

Figure 2b
PPG 425 in methanol; 150 °C; 200 atm/10 min; 200–350 atm/15 min; 350 atm – hold; 10 meter DB-17 column; Atn. = 4; Chart speed = 0.2 cm/min.

methanol solvent. This low molecular weight solvent elutes according to temperature differences ($tr_{100\,°C} > tr_{150\,°C}$). However, subsequent peaks which are due to the PPG oligomers elute based on the CO_2 density ($d_{CO_2\,-\,100\,°C} > d_{CO_2\,-\,150\,°C}$). The resolution of the peaks in Figure 2b is aided by the lower density values. Temperature again has an effect. The net result is better resolution in Figure 2b with comparable retention times to Figure 2a separations.

Reproducibility and repeatability data, as shown in **Table 1**, were also obtained for PPG 425 analyzed on different columns and at different times. The data reported for five of the most distinct peaks show the worst precision of only ± 2% for the areas and ± 1% for the retention times.

Table 1
Reproducibility data for PPG-425

	Run 1		Run 2		Run 3		Run 4	
Peak	R. T.	Area %	R. T.	Area %	R. T.	Area %	R. T.	Area %
A	9.06	10.25	9.09	9.99	30.97	9.62	30.98	9.79
B	9.76	17.84	9.86	17.93	31.93	17.68	31.94	17.79
C	10.82	30.73	11.04	30.53	33.11	30.52	33.92	30.51
D	12.04	25.75	12.34	25.85	34.36	26.11	34.36	25.99
E	13.27	15.44	13.62	15.70	35.62	16.07	35.67	15.92

Table 1, continued

	Run 5		Run 6		Run 7	
Peak	R. T.	Area %	R. T.	Area %	R. T.	Area %
A	22.36	9.46	22.38	9.33	22.35	9.50
B	23.37	17.24	23.38	17.10	23.36	17.34
C	24.62	29.94	24.62	29.87	24.60	30.08
D	25.95	26.45	25.94	26.55	25.92	26.43
E	27.32	16.91	27.29	17.14	27.27	16.64

Notes: 1. For each run, the peak with the largest area was designated as C; the two immediately preceding were designated A and B; and the two peaks following C were designated D and E.

2. Retention times are in minutes.

3. Chromatographic conditions:

Runs 1 and 2 (1986 October 27)
10 m DB-17 – 100 °C
200 atm/10 min, 200–300 atm/15 min

Runs 3 and 4 (1987 January 15)
20 m DB-225 – 150 °C
200 atm/10 min, 200–300 atm/25 min

Runs 5, 6, and 7 (1987 January 23)
20 m DB-225 – 150 °C
200 atm/10 min 200–350 atm/25 min

Statistical Analysis of Table 1

For the retention times, standard deviation calculations were made on data from Runs 5, 6, and 7.

Runs 5, 6, and 7

Peak	Avg. R. T.	Std. dev.
A	22.36	0.02
B	23.37	0.01
C	24.61	0.01
D	25.94	0.02
E	27.29	0.02

These data show that, for a given set of conditions, retention times are very reproducible.

Area percent data should be identical, regardless of the column or programming conditions used. Statistical analysis of the area percent data is listed below.

Peak	Avg. Area %	Std. Dev.
A	9.70	0.32
B	17.56	0.33
C	30.31	0.34
D	26.16	0.32
E	16.26	0.64

These values also show good reproducibility.

9.3.1.5.2 Polyethylene Glycols (PEG)

Polyethylene glycol oligomer mixture analyses were also studied extensively. A number of molecular weight ranges exist for polyethylene glycols. Separations of these oligomer mixtures can be achieved on different types of columns although DB-17 columns (10 m) tended to give the best separations. **Figure 3a** is a chromatogram of PEG of average molecular weight 190–210, separated on a DB-5 (10 m) column. **Figure 3b** shows the separation of PEG 380–400 on a DB-17 (10 m) column. When the same analysis is performed on a DB-225 (20 m) column, the resolution is very good, as shown in **Figure 3c**, but the retention times are prohibitively long. Attempts at separating the PEG 950–1050 with the 20 m DB-225 column were unsuccessful. When plugging of restrictors and splitters was not a problem, the retention of solvents was in some instances found to double. Partial plugging of the column restrictor caused the linear velocity to decrease, resulting in longer retention times. This problem was alleviated by repeated injection of solvent at high densities, which ultimately removed the partial clog. The separation of PEG 950–1050 was, however, achieved with a 10 m DB-17 column. Although, as **Figure 3d** shows, the resolution is good, the peaks tend to be broad.

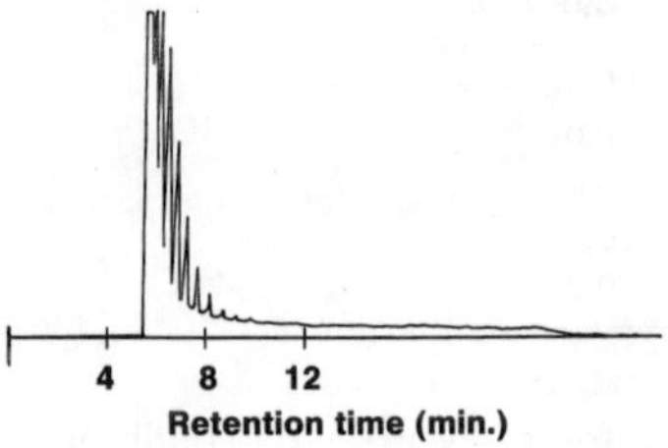

Figure 3a
PEG 190–210 in methanol; 150 °C; 180–280 atm/10 min; 280–300 atm/5 min; 300 atm – hold; 10 meter DB-5 column; Atn. = 4; Chart speed = 0.5 cm/min.

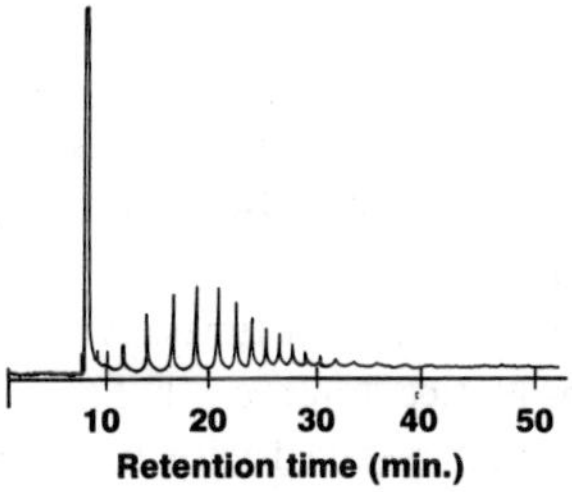

Figure 3b
PEG 380–400 in methanol; 150 °C; 200 atm/10 min; 150–350 atm/15 min; 350 atm – hold; 10 meter DB-17 column; Atn. = 4; Chart speed = 0.2 cm/min.

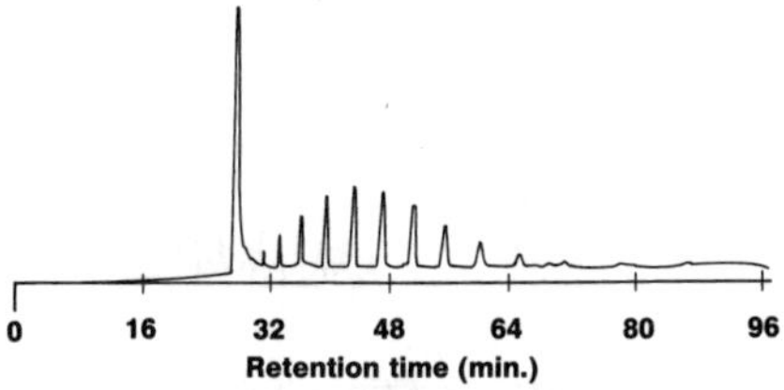

Figure 3c
PEG 380–400 25% in methanol; 150 °C; 200 atm/10 min; 200–350 atm/25 min; 350 atm – hold; 20 meter DB-225 column; Atn. = 3; Chart speed = 0.2 cm/min.

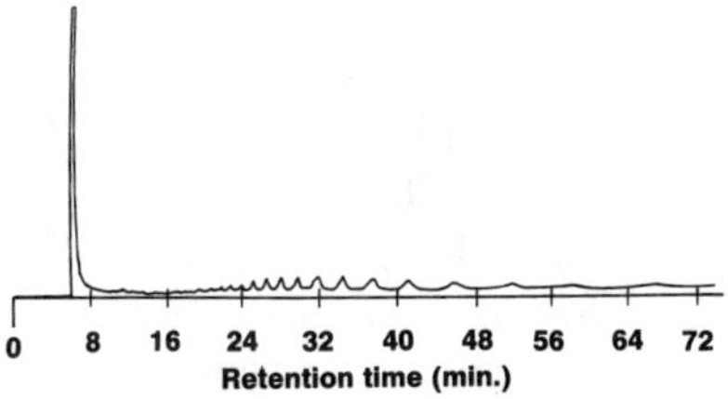

Figure 3d
PEG 950–1050 in methanol; 150 °C; 200 atm/10 min; 150–350 atm/15 min; 350 atm – hold; 10 meter DB-17 column; Atn. = 4; Chart speed = 0.2 cm/min.

9.3.1.5.3 Triton X-100

Triton X-100 is another oligomer mixture that is a good standard for SFC studies and column evaluations. As **Figures 4a–4d** show, this mixture can be separated on different type and size columns. Again, the issue becomes that of a compromise between resolution and length of retention times. A 10 m DB-17 column appears to be the compromise column. Notice, however, that the components intermediate between the major peaks are best resolved with the longer DB-225 (20 m) column (Figure 4d) and rapid analysis of the main components can be achieved with a 2-meter DB-5 column (Figure 4c).

After evaluation of chromatographic performance using standard materials, samples of interest to Alcoa were tested. These samples included high temperature poly(α-olefins) and lubricant materials.

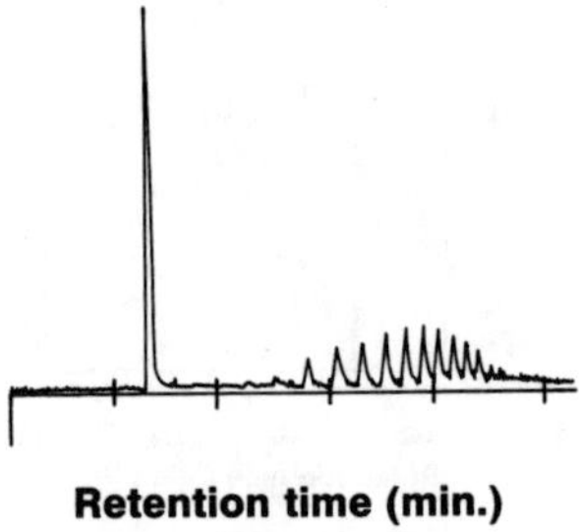

Figure 4a
Triton X-100 5% in methanol; 125 °C; 150 atm/15 min; 150–300 atm/10 min; 350 atm – hold; 10 meter DB-5 column; Atn. = 4; Chart speed = 0.2 cm/min.

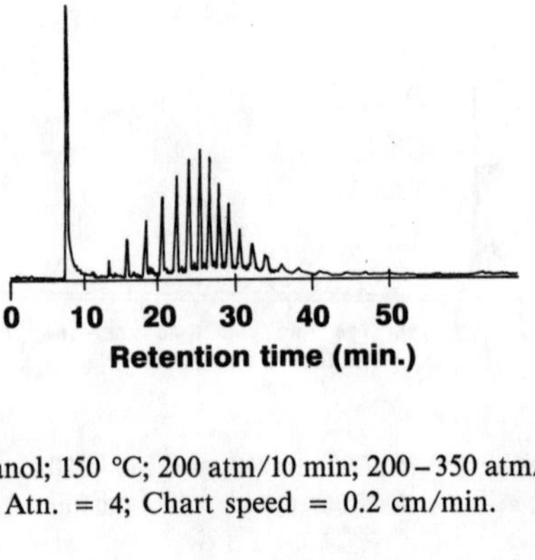

Figure 4b
Triton X-100 5% in methanol; 150 °C; 200 atm/10 min; 200–350 atm/15 min; 350 atm – hold; 10 meter DB-17 column; Atn. = 4; Chart speed = 0.2 cm/min.

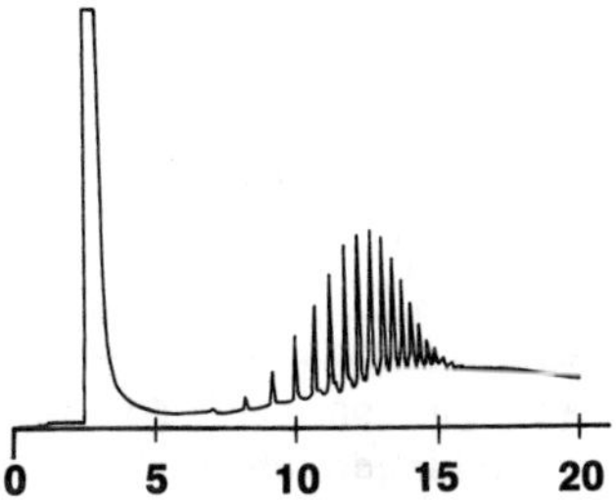

Figure 4c
Triton X-100 5% in methanol; 150 °C; 150 atm/5 min; 150–350 atm/10 min; 350 atm – hold; 2 meter DB-5 column; Atn. = 2; Chart speed = 0.5 cm/min.

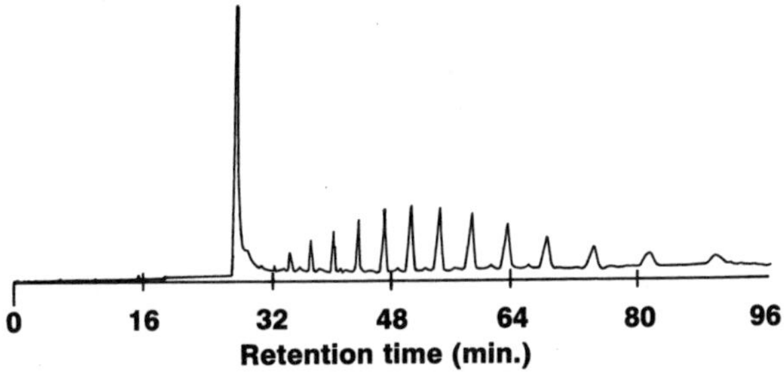

Figure 4d
Triton X-100 5% methanol; 150 °C; 200 atm/10 min; 200–350 atm/25 min; 350 atm – hold; 20 meter DB-225 column; Atn. = 3; Chart speed = 0.2 cm/min.

9.3.1.5.4 Lubricant Materials

Lubricant A is a lubricant whose composition is considered proprietary by Alcoa. Its constituents are difficult to characterize by conventional chromatographic and/or other techniques. The work done so far shows that SFC can indeed be used as a qualitative analytical technique with the potential to perform quantitative assays.

Figure 5a shows a chromatogram of the new Lubricant A (50% in hexane) and **Figure 5b**, those of the used Lubricant A sampled at different times. The important observation is the ability of the technique to track the depletion of the lubricant components with usage. This can be an important tool in evaluating the relative importance of each of the ingredients by monitoring the levels of these components

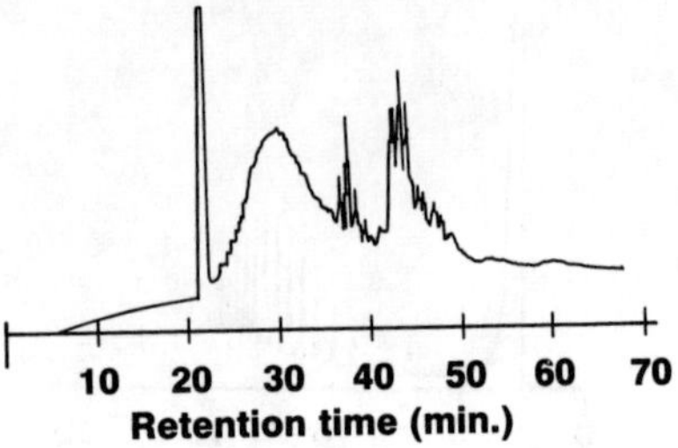

Figure 5 a
New lubricant in hexane; 100 °C; 100 atm/5 min; 100 – 255 atm/25 min; 255 – 350 atm/60 min; 20 meter DB-225 column; Atn. = 4; Chart speed = 0.2 cm/min.

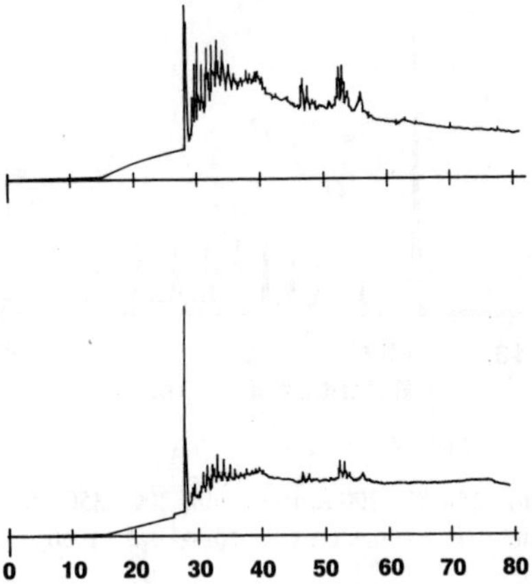

Figure 5 b
A) Used lubricant sample 1; B) Used lubricant sample 2; 100 °C; 100 atm/5 min; 100 – 255 atm/25 min; 255 – 350 atm/60 min; 20 meter DB-225 column; Atn. = 4; Chart speed.

and comparing them to the performance of the lubricant. This can provide information on whether the replenishment of one additive can actually extend the life of the lubricant. The individual components were also analyzed. As shown in **Figures 5 c and 5 d**, most of these additives can be easily resolved from the base oil component, which is the broad peak after the solvent in Figure 5 a. Some of the Additive D components are, however, masked by the other constituents.

Figure 5 e, comparative chromatograms of new and used lubricant on larger diameter DB-WAX, show that although resolution among components is poor,

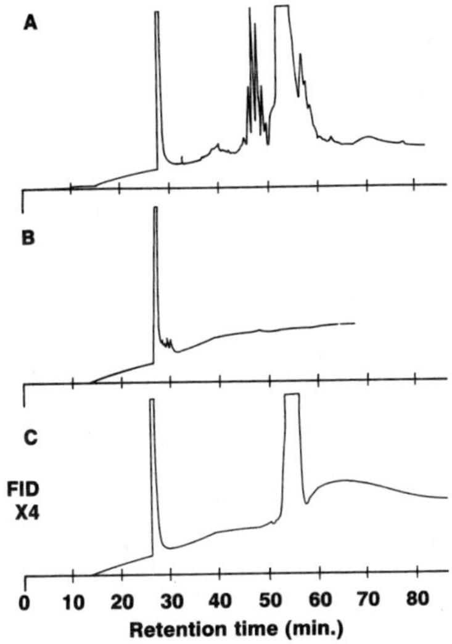

Figure 5c
A) Additive A; B) additive B; C) additive C; 100 °C; 100 atm/5 min; 100 – 255 atm/25 min; 255 – 350 atm/60 min; 20 meter DB-225 column; Atn. = 4; Chart speed = 0.2 cm/min.

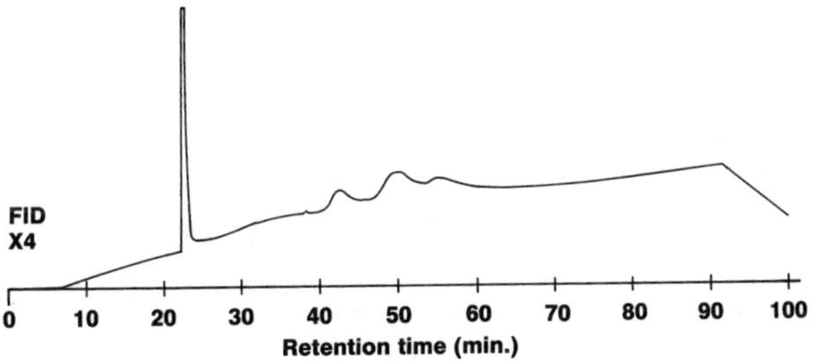

Figure 5d
Additive D; 100 °C; 100 atm/7 min; 100 – 255 atm/25 min; 255 – 350 atm/60 min; 20 meter DB-225 column; Atn. = 4; Chart speed = 0.2 cm/min.

evaluation of total composition can be done faster than on smaller diameter DB-225 columns, as shown before in Figures 5a and 5b. Efforts to better resolve these materials and detect lower levels, especially with packed columns, are continuing.

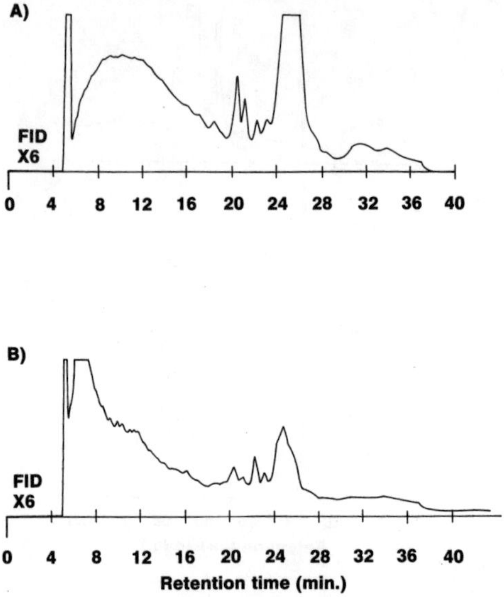

Figure 5e
Comparative Chromatograms of new (A) and used (B) lubricant on a 2 meter DB-WAX column; 150 °C; 0.16 g/ml/6 min; 0.16–0.28 g/ml/7 min; 0.28–0.52 g/ml/7 min; 0.52–0.58 g/ml/4 min; 0.58 g/ml/3 min; 0.58–0.66 g/ml/3 min.

9.3.1.5.5 Poly(α-olefins)

Poly(α-olefins) (PAOs) with branched C_{10}-repeating units can be used in place of straight chain hydrocarbons for particular applications. The specification of properties is usually by viscosity. This, however, does not provide information on the content of the low molecular weight *vs.* the high molecular weight compound composition. SFC can be applied to determine molecular weight distributions which give tighter specifications on the product. **Figure 6a** shows comparative analyses of poly(α-olefins) from three manufacturers. The results indicate different levels of C_{30}, C_{40}, and C_{50} moieties for the resolved peaks and unresolved C_{60} and C_{70}, respectively, in the three samples. Thus, even though the viscosity may be similar, the performance of the material may ultimately depend on the relative concentrations of the oligomers. **Figure 6b** also shows separations of PAOs of average molecular weight of 400 and 1000. Although the chromatography is not really good, comparisons can be made between different formulations.

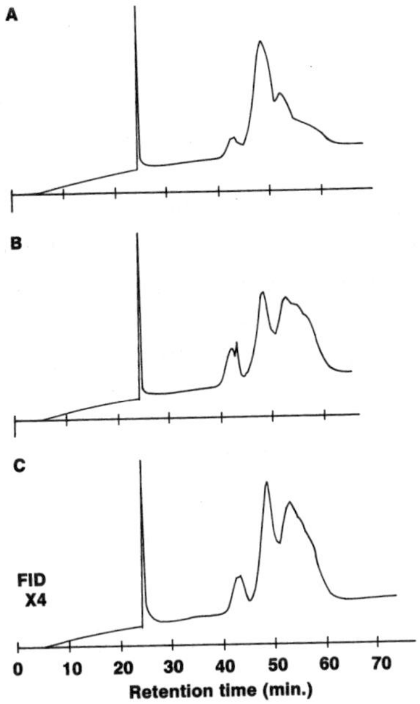

Figure 6 a
A) Emery 3008; B) Mobil SHF82; C) Chevron Oronite; 100 °C; 80 atm/5 min; 80 – 300 atm/60 min; 300 atm – hold; 20 meter DB-225 column; Atn. = 4; Chart speed = 0.2 cm/min.

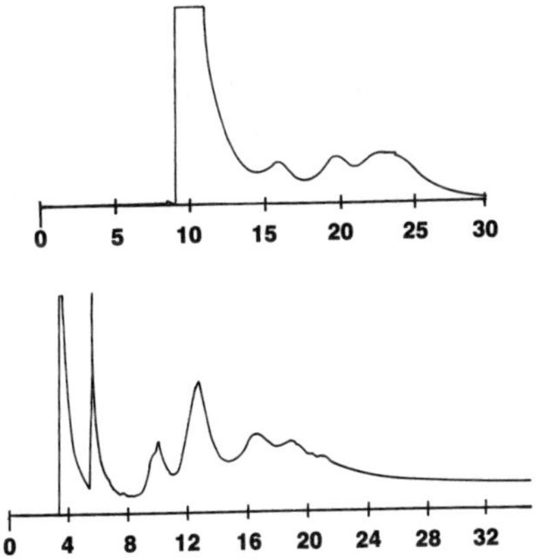

Figure 6 b
PAO 400 in hexane; 100 °C; 150 – 250 atm/10 min; 250 atm hold/5 min; 2 meter DB-5 column; Atn. = 4; Chart speed = 0.5 cm/min.

9.3.1.5.6 Fish Oils

The increasing interest in the use use of fish oils as a dietary supplement for combating cholesterol buildup in blood has resulted in an eager search for the active constituents of those oils. The cholesterol combating agents are suspected to be omega-3 fatty acids. Extraction and purification of these components has not been easy due to the presence of other acids and esters in fish oils.

Figure 7a depicts a chromatogram of fish oils. The elution of compounds towards the end of the run, suspected to be triglycerides, indicates that the oils components can be resolved by SFC. Identification of those components was not possible. A mass spectroscopic detector would be required for such an identification.

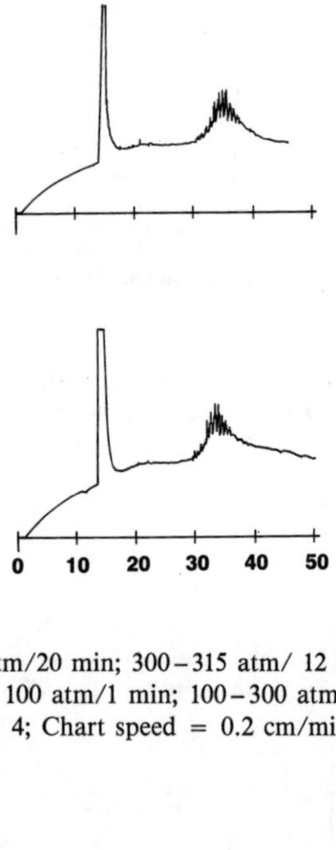

Figure 7a
A) 100 atm/1 min; 100–300 atm/20 min; 300–315 atm/ 12 min; 315 hold/ 20 min.
B) Fish oil in hexane; 150 °C; 100 atm/1 min; 100–300 atm/20 min; 300–350/40 min; 20 meter DB-225 column; Atn. = 4; Chart speed = 0.2 cm/min.

9.3.2 Packed Column SFC

The use of packed column SFC was explored due to curiosity and as an effort to alleviate the problems encountered with capillary SFC. The curiosity was to determine if a column (10 cm x 1 mm i.d.) packed with neutral alumina could be used for the separations, as explained above for capillary columns.

The first column used was packed with neutral alumina (5 μm particle size). **Figure 8a** shows two chromatograms obtained with two hydrocarbon standards. The main features of these chromatograms are the baseline resolution of the peaks, the minimal peak tailing and the very short retention times. Similar GC analyses would require either longer retention times or more severe temperature conditions than the 90 °C used here.

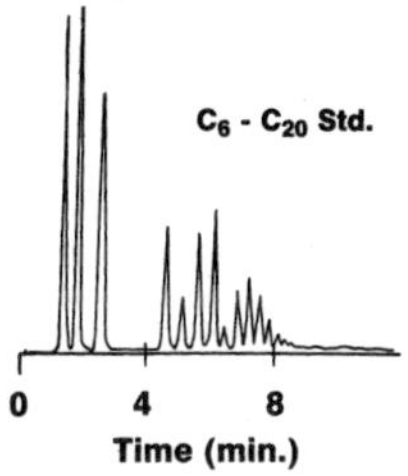

Figure 8a
Chromatograms of hydrocarbon standards obtained on a normal phase alumina column; 90 °C; 0.408 g/ml hold/2 min; 0.408–0.670 g/ml/6 min; 0.670–0.784 g/ml/6 min; 0.784 g/ml hold/1 min; 0.784–0.408 g/ml/4 min.

Separations of the poly(α-olefins) achieved earlier on the fused silica columns were also attempted with the packed columns. **Figures 9a, 9b** and **9c** depict normal phase alumina separations of various poly(α-olefins) supplied by different vendors. Besides providing comparative molecular weight distributions, individual oligomer separations can also be achieved.

One potential application of alumina packed column SFC is simulated distillation analysis. This technique, which provides information on the distribution of compounds in a mixture by boiling point, is widely used in the crude oil and petroleum industries in place of manual distillation techniques. The gas chromatographic technique currently used, however, requires programming to high temperatures (approx. 430 °C) in order to elute the high boiling point compounds. Besides the fact that some of these compounds may not be eluted, there is also a likelihood that the high temperatures may "pyrolyze" some of the more labile highly branched hydrocarbons. SFC with packed colums offers an alternative, providing low temperature separations and achieving the resolution needed to perform the analysis.

Figures 10a and **10b** depict chromatograms obtained for a C_{12}- to C_{44}-hydrocarbon standard and a polyethylene (av. M. W. = 740) oligomeric mixture, respectively. The success of these separations is shown by the baseline resolved peaks, the

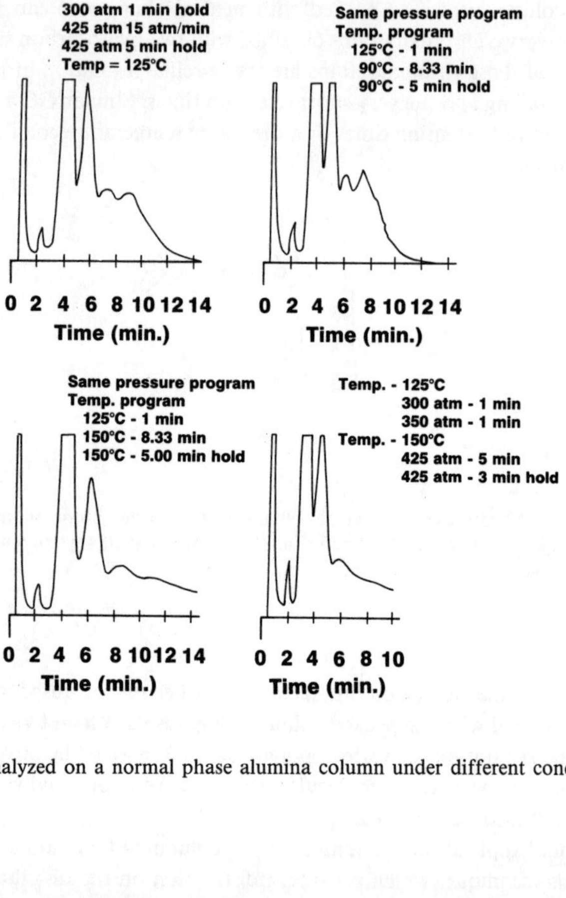

Figure 9a
Emery 3008 analyzed on a normal phase alumina column under different conditions.

small tailing in peaks and the short retention times. These separations were also achieved at temperatures of 125 °C or less.

Other columns packed a with functionalized alumina have been tried for SFC analysis. **Figure 11a** shows a separation on a C_{18}-alumina column of the same C_6–C_{20} standard as in **Figure 8a**. Although the resolution is better with the C_{18}-column, there is some peak tailing and longer retention times are observed. Isomer separation might be achieved better with the C_{18}-column. **Figure 11b** shows a comparative separation of a poly(α-olefin) (av. M. W. = 400) on a normal phase and a C_{18}-alumina column. Better resolution of this material is obtained with the latter. Separations of hydrocarbon standards were also tried on aminophenyl- and C_8-alumina columns. Again (as shown in **Figures 11c and 11d**, respectively), good component resolution is obtained with both columns at reasonable retention times.

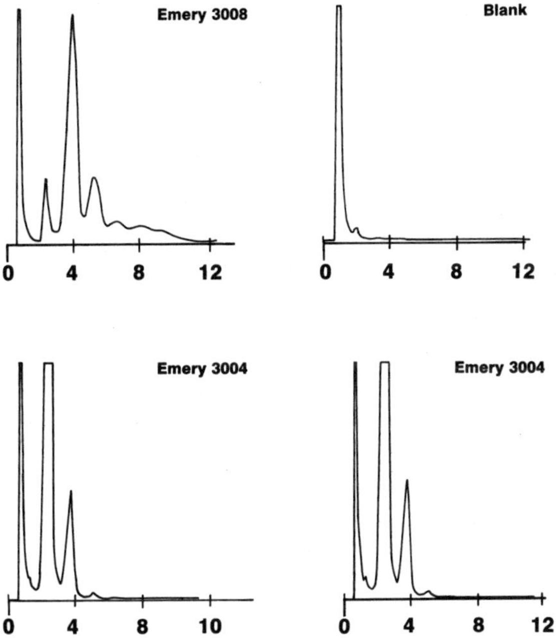

Figure 9 b
Chromatograms of Emery 3008 and Emery 3004 obtained on a normal phase alumina column. 125 °C; 300 atm/1 min; 425 atm/15 atm/min; 425 atm/2 min hold.

There is an increasing interest in polymer-coated alumina and silica as packings for liquid chromatographic separations. In this laboratory, SFC separations were attempted with polybutadiene (PBD)-coated alumina (10 μm) particles.

As **Figures 11 e** and **11 f** show, baseline resolution of hydrocarbon mixtures can be obtained with 10 μm-PBD coated alumina packings at retention times similar to those obtained with uncoated and C_{18}-alumina packings (Figures 8 a and 11 a). The latter packings were made from 5 μm-size alumina particles. Figure 11 g, a chromatogram of a shale oil extract, shows that alkene-alkane hydrocarbon pairs can be resolved.

The capability of those large particle sizes to provide such resolving capabilities is important for preparative or process scale separations. The lower pressures required for separations with such particles ensure that analytical separation can be easily scaled up.

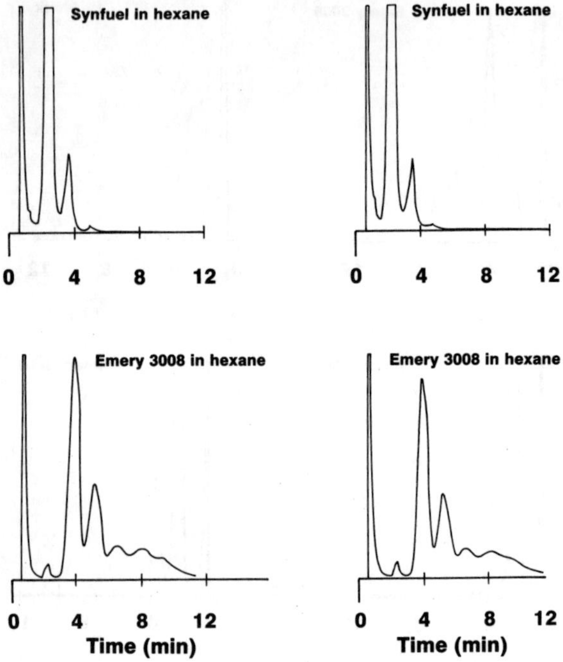

Figure 9 c
Comparative chromatograms of Chevron Synfuel and Emery 3008 analyzed on a normal phase column Program; 125 °C – 300 atm/1 min; 90 °C – 425 atm/15 atm/min; 90 °C – 425 atm/3 min.

The polymer-coated surface of the alumina particles should be inert enough to allow SFC of moderately polar compounds with supercritical carbon dioxide as the mobile phase. These applications are currently being evaluated.

The above account describes the applications of alumina packed SFC columns for hydrocarbon-based materials. Work is in progress to develop applications for polar or compounds other than hydrocarbons. This effort is two-fold: (a) test functionalized aluminas (C_{18}, C_8, phenyl-, *etc.*) for efficiency of separation of non-hydrocarbon compounds, and (b) develop functionalized aluminas and polymer-coated aluminas that are targeted towards specific SFC analyses.

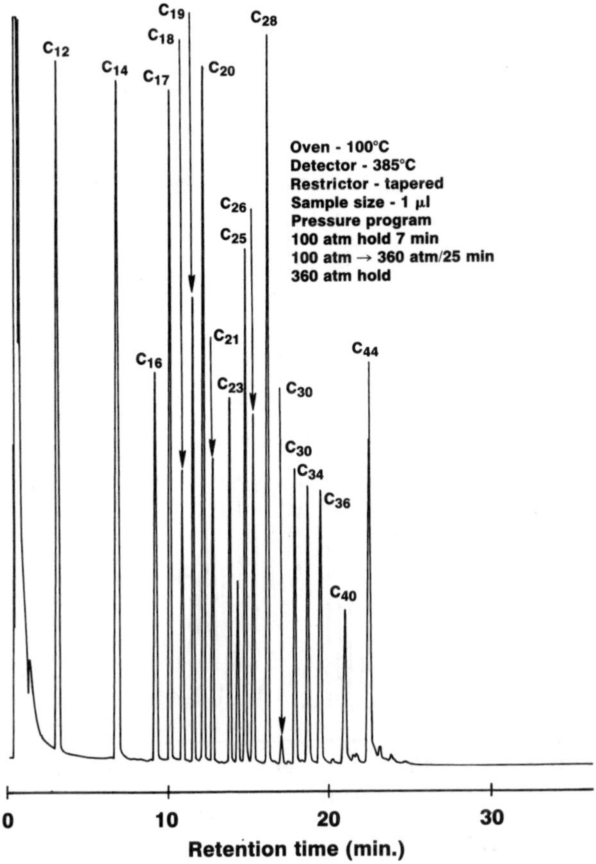

Figure 10a
$C_{12}-C_{44}$ hydrocarbon standard analyzed on a normal phase alumina column.

9.3.3 Applied Separations

The modeling of supercritical fluid desorption based separations can be achieved by using the simultaneous extraction or desorption/analysis unit. The example cited here is the removal of impurities from a high molecular weight polyester using supercritical fluid carbon dioxide. A one gram sample was used for the test.

Figure 12a shows the chromatogram of the real time total extractives eluted from the reactor. This shows that within 40 minutes, most of the extractable materials at

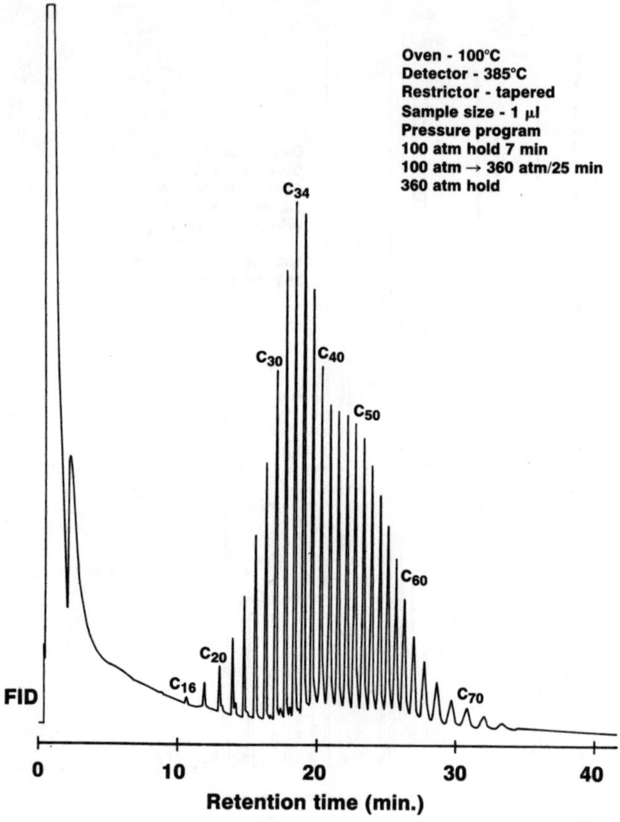

Figure 10 b
PE 740 standard analyzed on a normal phase column.

this temperature and pressure (80 °C and 150 atm.) had been removed. For example, with an exit CO_2 flow rate of 80 ml/min, the mobile phase requirements are about 3.2 liters of gaseous CO_2/gram of polyester. The samples obtained at approximately 10-minute intervals **(Figure 12 b)** indicate the elution of one major compound that is likely to be a residual monomer or solvent.

By increasing the pressure to 175 atm, hence, increasing density, additional materials are eluted. The analyzed extractives **(Figure 12 c)** indicate the elution of an oligomeric type material suspected to be low molecular weight products of side reactions or prepolymer mixtures. Again, these impurities are quickly removed, as shown by the successive analyses.

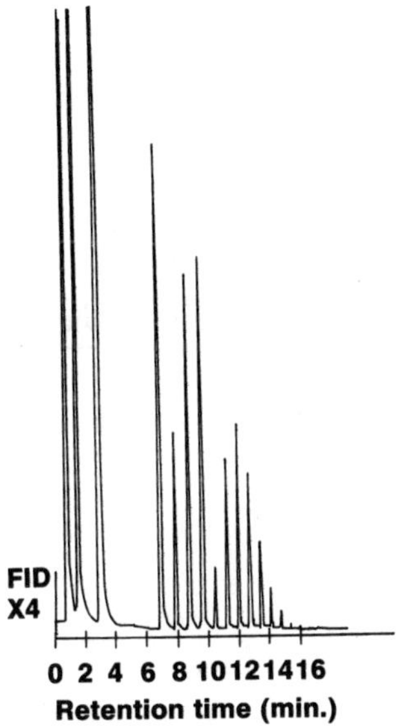

Figure 11 a
C_6-C_{20} hydrocarbon standard on a C_{18} alumina column.

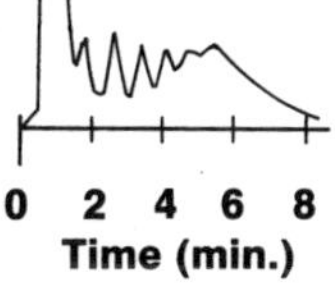

Figure 11 b
PAO 400 analyzed on normal phase (NP) – and C_{18} alumina columns. Program: 110 °C – 80 °C at 300 atm; 80 °C – 300 – 400 atm at 35 atm/min; 400 atm/3 min hold.

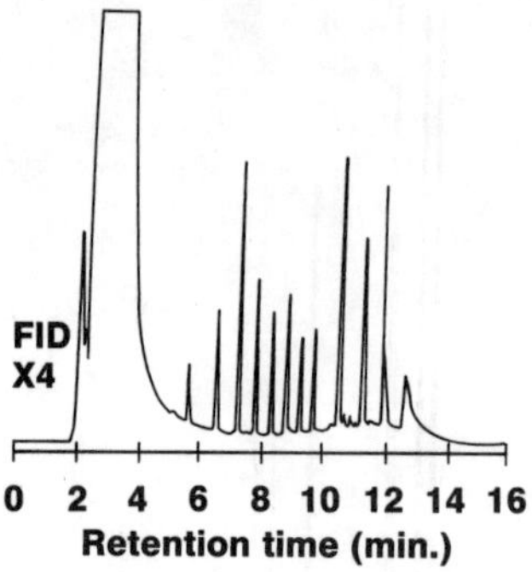

Figure 11c
Hydrocarbon std. analyzed on an aminophenyl column. 110 °C; 100 atm/3 min; 100–390 atm/8 min; 390–450 atm/1.5 min; 450 atm/3 min hold.

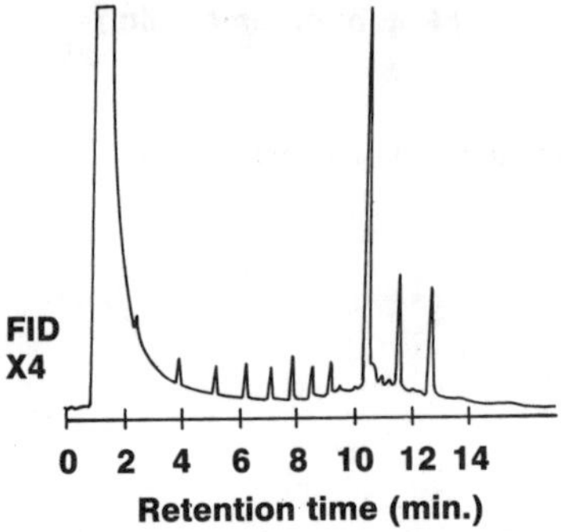

Figure 11d
Sample = hydrocarbon mixture. Conditions: 110 °C; 150 atm/2 min; 150–350 atm 20 atm/min; 350 atm/3 min hold.

The bottom line is that this polyester material can be purified rapidly by using SFC conditions that are optimum for removing the impurities but not severe enough to dissolve or attack the polyester itself.

The same experiment can be performed for the solid extraction of coffee beans, tea leaves, lard, gums or for desorption studies of materials preadsorbed onto adsorbents such as silica, alumina, titania, zirconia, *etc.*

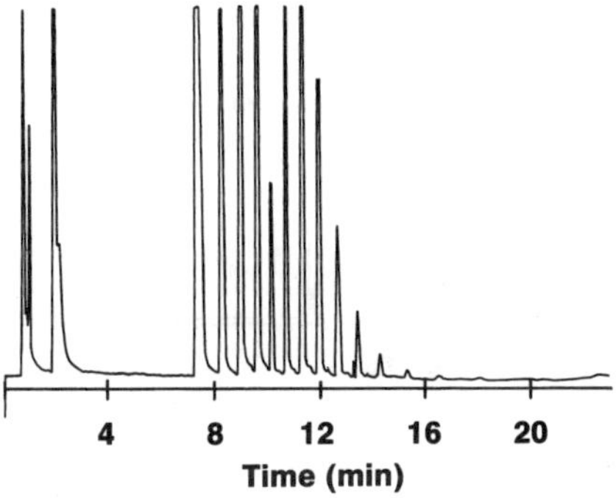

Figure 11e
C_6-C_{20} hydrocarbon std. analyzed on polymer (PBD) coated alumina particles (10 μm). Conditions: Oven temp. 80 °C. Program: 150–150 atm/5 min; 150–250 atm/5 min; 250–275 atm/4 min; 275–275 atm/6 min; 275–150 atm/3 min.

9.4 Conclusions

The work performed so far indicates that supercritical fluid chromatography is a viable analytical technique that can also find applications in modeling process scale separations.

The analytical separations ranged from non-polar hydrocarbons to more complex mixtures of hydrocarbons and polar acids or esters. SFC can indeed be used to characterize low molecular weight polymer oligomer mixtures.

The applications developed confirm the practical applications of both capillary and packed column SFC. The non-polar surfaces achieved with crosslinked or bonded capillary columns provide excellent separations of compounds. Packed columns, on the other hand, provide fast analyses with very good resolution and show potential for quantitative analyses to low levels. The drawback with packed columns is still the incapability of performing polar compound separations, due to active sites on most packing particles that are present even on the functionalized substrates when using pure CO_2. Polymer-coated alumina particles are currently being explored for SFC applications of polar compounds.

The work done with the SFE/SFC unit demonstrates the potential for modeling process scale separations using small bench scale equipment. Application areas include the purification or separation of food, cosmetic and pharmaceutical products. The more corrosive supercritical fluids such as supercritical water can be used to purify materials, such as silica, which dissolve in this fluid [20] and can be precipitated away from the original matrix.

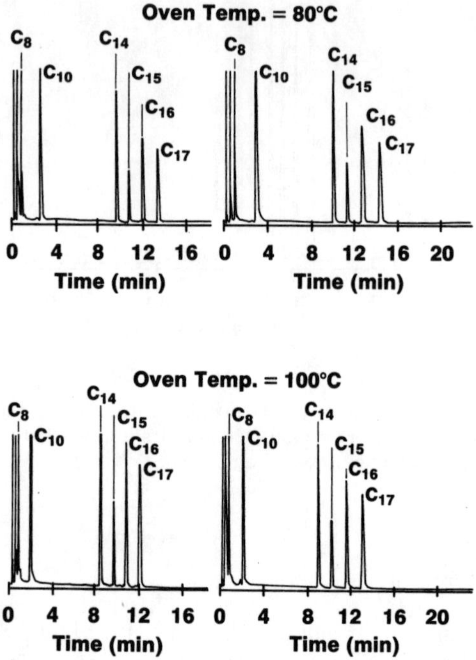

Figure 11f
C_6-C_{17} hydrocarbon std. mixture analyzed on a PBD-coated alumina (10 μm) column. Conditions: 150–150 atm/5 min; 150–250 atm/5 min; 250–275 atm/4 min; 275–275 atm/6 min; 275–350 atm/5 min; 350–350 atm/5 min; 350–150 atm/5 min.

The purification of polymers is another area that will find increasing supercritical fluid applications. The removal of solvents, for example, is critical in polymers that are or will be used for drug delivery or food packaging materials. The use of supercritical carbon dioxide, a non-toxic gas obtainable at high purity levels, would be ideal for such purposes.

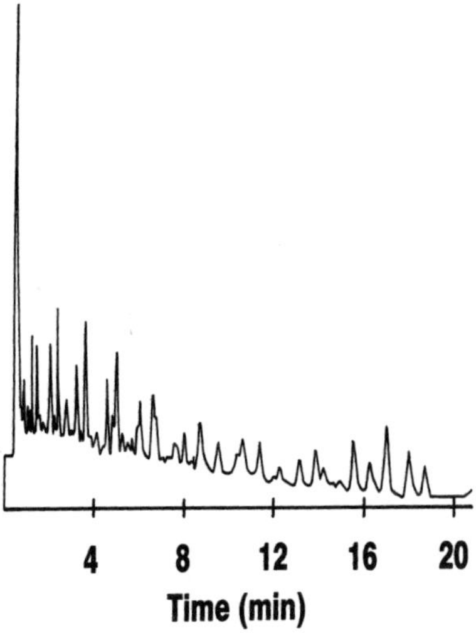

Figure 11g
Shale oil (pentane extract) analyzed on a PBD-coated alumina column (10 μm particles). 150–150 atm/5 min; 150–250 atm/5 min; 250–275 atm/4 min; 275–275 atm/6 min; 275–350 atm/5 min; 350–350 atm/5 min; 350–150 atm/5 min; oven temp. 80 °C.

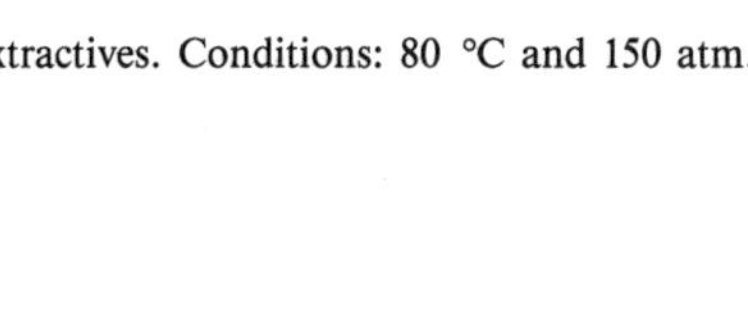

Figure 12a
Chromatogram of total extractives. Conditions: 80 °C and 150 atm.

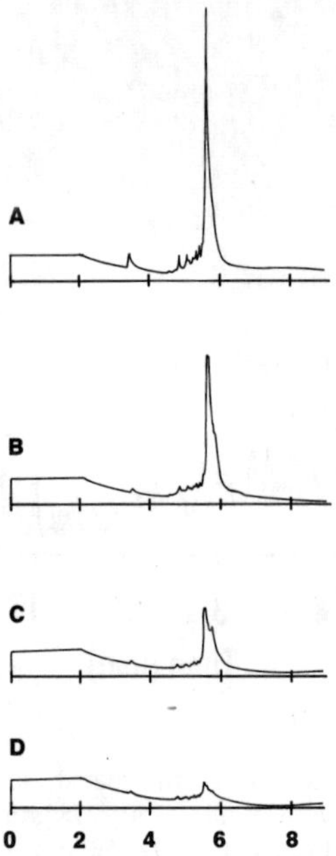

Figure 12 b
SFC of extractives. Conditions: 110 °C at 350 atm

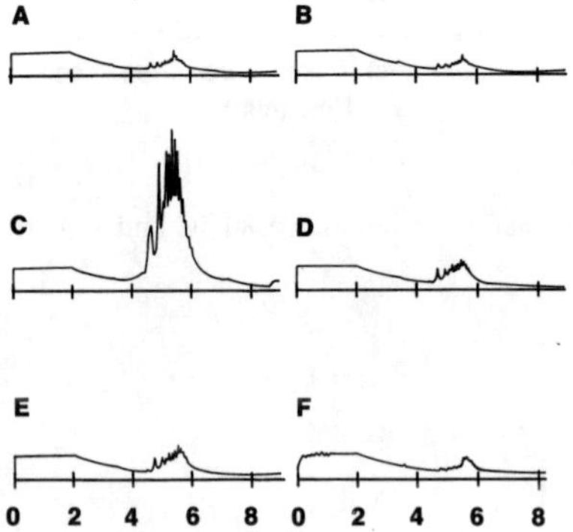

Figure 12 c
SFC of extractives. Conditions: 110 °C at 350 atm.

Acknowledgments

This work would not have been possible without the expert technical assistance of *Douglas G. Weaver, Karen Fraser-Mills,* and *Richard Burr* of Alcoa and *Harold A. Bryant* of the Chemistry Department at the University of Pittsburgh. The support of *Dr. Larry Wieserman* and *Dr. Lin*, of Alcoa, by providing the functionalized and polymer-coated alumina is appreciated. I would also like to give due credit to *Dr. M. Lalit Kumar* of Suprex Corporation for awakening interest in SFC and the applications group at Suprex for their assistance.

References

[1] *C. M. White* and *R. K. Houck*, HRC & CC **9** (1986) 4–17.
[2] *M. McHugh* and *V. Krukonis*, "Supercritical Fluid Extraction – Principles and Practice," Butterworths (1986) p. 13.
[3] *Ibid*, p. 18.
[4] *E. Klesper, A. H. Corwin*, and *D. A. Turner, J. Org. Chem.* **27**(1962) 700–701.
[5] *E. Klesper* and *W. Hartmann,* Polymer Letters Edition **15** (1977) 707–712.
[6] *S. T. Sie, W. van Beersum,* and *G. W. A. Rijnders,* Sep. Sci. **1** (1966) 459–490.
[7] *S. T. Sie* and *G. W. A. Rijnders,* Sep. Sci. **2** (1967) 699–727.
[8] *S. T. Sie* and *G. W. A. Rijnders* Sep. Sci. **2**729–753
[9] *S. T. Sie* and *G. W. A. Rijnders,* Sep. Sci. **2** (1967) 755–777.
[10] *S. T. Sie* and *G. W. A. Rijnders,* Anal. Chim. Acta **38** (1967) 31–44.
[11] *S. T. Sie, J. P. A. Bleumer,* and *G. W. A. Rijnders,* "Gas Chromatography 1968," *C. L. A. Harbourn* and *R. Stock* (eds), Elsevier, Amsterdam (1969).
[12] *L. McLaren, M. N. Myers,* and *J. C. Giddings,* Science **159** (1968) 197–199.
[13] *J. C. Giddings, M. N. Myers, L. McLaren,* and *R. A. Keller,* Science **162** (1968) 67–73.
[14] *D. R. Gere,* Science **222** (1983) 253–259.
[15] *C. M. White* and *R. K. Houck,* Technical Bulletin # 201, Suprex Corporation (1985).
[16] *C. M. White* and *R. K. Houck,* HRC & CC **8** (1985) 293–296.
[17] Technical Bulletin # 87–4,1 Suprex Corporation (1987).
[18] Technical Bulletin # 87-APPL, Suprex Corporation (1987).
[19] *M. McHugh* and *V. Krukonis,* "Supercritical Fluid Extraction, Principles and Practice," Butterworths (1986).
[20] *M. McHugh* and *V. Krukonis. ibid,* p. 20.

Acknowledgment

References

10 Restrictor Performance Characteristics for SFC

Bob W. Wright and Richard D. Smith
Chemical Sciences Department, Pacific Northwest Laboratory[1)], Richland, WA 99352, USA

Dr. Bob W. Wright is a Senior Research Scientist in the Chemical Sciences Department of Battelle, Pacific Northwest Laboratories. He obtained his Ph. D. in Chemistry from Brigham Young University in 1982 after which he joined the Battelle staff. His primary research interests include the development and application of high resolution separation techniques such as capillary gas chromatography, supercritical fluid chromatography, and capillary zone electrophoresis. He also has interests in fundamental solvation phenomena and in molecular interactions with chromatographic surfaces. He has also been involved in the development of combined supercritical fluid chromatography-mass spectrometry and analytical supercritical fluid extraction methods. *Dr. Wright* is the author of more than sixty publications.

Dr. Richard D. Smith is a Senior Staff Scientist and Group Leader of the Chemical Methods and Separations Group in the Chemical Sciences Department of Battelle, Pacific Northwest Laboratories. *Dr. Smith* obtained his Ph. D. at the University of Utah in 1975 in Physical Chemistry. He spent a year as an NRC postdoctoral associate at the Naval Research Laboratory in Waschington, DC, before joining Battelle in 1976. *Dr. Smith's* research has involved the development and application of new separation methods and analytical techniques with an emphasis

[1)] Operated by Battelle Memorial Institute for the U. S. Department of Energy under Contract DE-AC06-76RLO 1830.

on combined methods using mass spectrometry. Related areas of research activity include studies of supercritical fluid solvation and colloidal phenomena. He has been involved in LC-MS, SFC-MS, and most recently, capillary zone electrophoresis-MS development. *Dr. Smith* is the author of over one hundred and thirty publications.

Summary:

Many of the detectors currently used for supercritical fluid chromatography (SFC) require the chromatographic effluent to be depressurized before detection. This is generally achieved by using some form of capillary restrictor. The performance of the restrictor and the efficiency for transfer of less volatile analytes to the gas phase is crucial for successful detection. An improved understanding of restrictor performance can be obtained by considering the thermodynamics involved during the expansion of the fluid. To maintain good flow characteristics and avoid a two-phase (condensed) fluid mixture during expansion, it is necessary for the fluid to have sufficient enthalpy. The success of transporting analytes to the detector also depends on the volatilities and fluid phase solubilities of the analytes, the restrictor design, and fluid pressure and temperature. For less volatile compounds, it is possible for analytes to precipitate and collect on the restrictor walls, which can lead to spiking and eventual plugging of the restrictor. Use of a restrictor design that maximizes fluid density at the restrictor exit minimizes this behavior. When analyte nucleation is delayed to near the end of the restrictor, smaller particles are produced and more efficient detection is obtained. Increased pressure serves to enhance transport of analytes as higher solubilities are always obtained, but increased temperature may be counterproductive due to lower fluid phase solubilities. For analytes exhibiting some volatility, however, higher temperatures can increase solubility. In a qualitative performance comparison of four restrictor types (straight-walled capillary, tapered, polished, and porous frit) using a polymer of relatively low volatility as test analytes, the tapered and polished restrictors more efficiently transported the higher-molecular-weight and less-volatile oligomers.

Concentrate:

The major consideration in selecting a restrictor should be the volatility of the analytes. Less volatile analytes will be more efficiently transported through a restrictor with a low length-to-diameter ratio. This allows an incomplete pressure drop to

occur over the restrictor, which maximizes the fluid density at the restrictor exit and minimizes analyte nucleation in the restrictor. Operation at higher flow rates also improves detection of less volatile analytes. For analytes volatile at restrictor temperatures, the most rugged and conveniently replaced restrictor is desirable.

10.1 Introduction

Many of the detectors currently used for supercritical fluid chromatography (SFC) such as flame ionization and mass spectrometry, require the chromatographic effluent to be depressurized before detection. Depressurization is generally achieved by expansion of the mobile phase through a restrictor into a region where detection can occur. Since proper conditions produce a gas after expansion, conventional gas-phase detection including electron impact (EI) and chemical ionization (CI) mass detection methods, including spectrometry are readily adapted. At present, most SFC utilizes some form of capillary restrictor with the dimensions empirically selected to give the desired mobile phase linear velocity for the chromatographic temperature and pressure range. The capillary restrictor can be short, with a relatively small inner diameter or longer with a larger i.d. The ideal restrictor for SFC would provide the following characteristics: (a) provide for uniform pulse-free flow; (b) offer immunity from plugging; (c) be easily replaced or provide for variation of flow rate and; (d) provide for complete transfer of labile or nonvolatile solutes to the detector without pyrolysis or formation of analyte particles. A wide variety of practical restrictors have been studied to date with each design having distinct advantages and disadvantages. However, it is clear that the ideal restrictor has yet to be described.

Since restrictor performance is vital for successful SFC and SFC-MS operation, this topic will be discussed in detail in this chapter. Although flow through even a simple capillary restrictor with subsequent expansion is an extremely complex process, significant insight into the performance of various restrictors can be obtained by considering the thermodynamic and kinetic processes occurring during expansion. The properties of analytes relevant to the expansion process will also be discussed as will the practical aspects of various restrictor designs. A performance comparison of four different restrictor types (straight-walled capillary, tapered or drawn, polished or integral, and porous frit) is described using flame ionization detection of a polyol polymer of relatively low volatility.

10.2 Thermodynamics of Fluid Expansion Through Restrictors

The characteristics of an SFC restrictor and, to a significant degree, the efficiency for transfer of less volatile solutes to the gas phase during the expansion, depend upon the state of the fluid prior to expansion, the dimensions of the restrictor, restrictor heating (and heat transfer properties of the restrictor), and the state of the fluid during and after the expansion. For this discussion, the simpler case of a straight-walled capillary or linear restrictor is addressed. However, similar (but more complex) considerations apply to other restrictor types that are currently more frequently used. **Figure 1** schematically illustrates an adiabatic process where the enthalpy (H) of the fluid is the same before and after the expansion process. Expansions relevant to SFC occur under conditions where the Joule-Thomson coefficient, $(\delta T/\delta P)_H$, is generally positive and net cooling results [1]. If the restrictor is very short the process may be considered adiabatic. In this case, the equilibrium state of the fluid after expansion (if also isolated from the surroundings) can be predicted directly from thermodynamic data. However, the state of the fluid during and shortly after expansion depends upon the physical processes related to the expansion that are kinetically controlled and inherently nonequilibrium. A detailed discussion of the thermodynamics of the expansion has been given elsewhere [2] and will be summarized here.

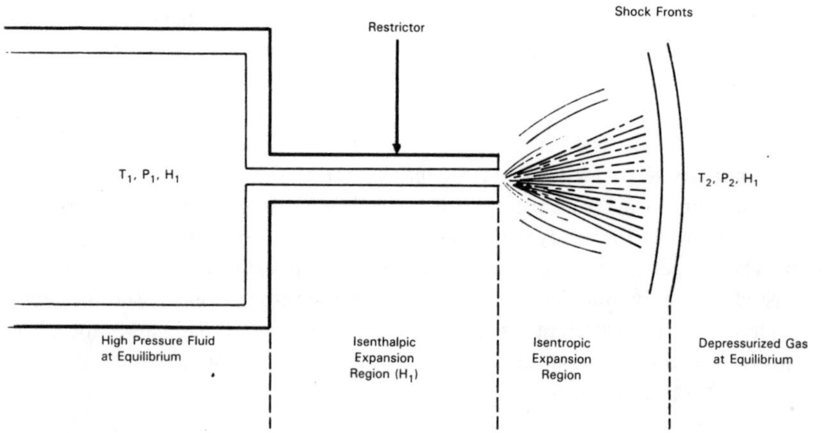

Figure 1
Schematic illustration of the supercritical fluid expansion of an adiabatic process where the initial and final states of the fluid are known. Reprinted with permission from Ref. [2] Copyright 1986, American Chemical Society.

Flow through a short restrictor will be nearly isenthalpic until the end of the restrictor, where isentropic expansion to supersonic velocities occurs. Shock phenomena and interactions with the background gas in the expansion region largely convert the kinetic energy of the gas jet back to internal energy [3]. For CO_2, an initial enthalpy of less than 175 kcal/g results in the expanding fluid traversing the two-phase region (often manifested by a build-up of frozen CO_2 at the restrictor exit). In order to obtain good flow characteristics and avoid a two-phase region during CO_2 expansion, a fluid temperature of at least 80 to 100 °C before entering the restrictor should be maintained. When an expansion occurs from pressures and temperatures that result in two phases, the assumption of adiabatic conditions allows the fraction of fluid in each phase to be calculated. **Figure 2** gives the fraction

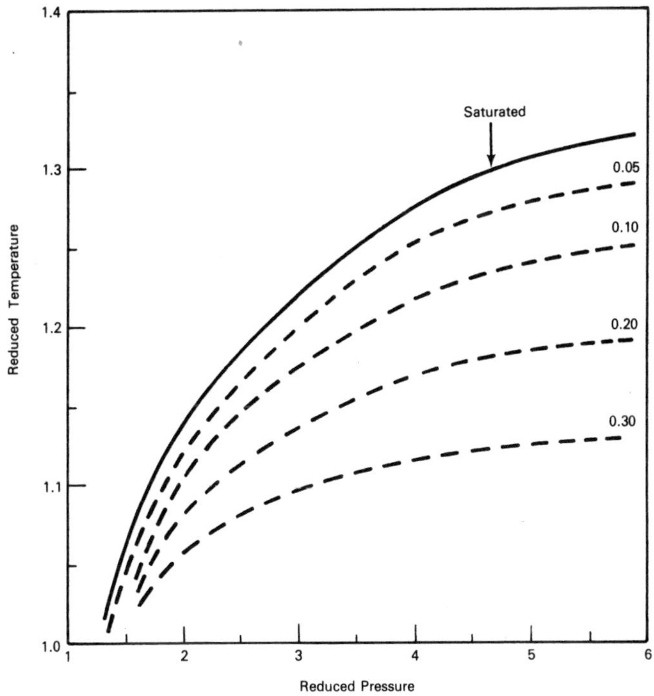

Figure 2
Reduced temperature-reduced pressure diagram for supercritical carbon dioxide showing constant vapor-liquid composition (dashed lines) after expansion to low pressures and the minimum pressure and temperature conditions (solid line) necessary to avoid condensation during expansion. Reprinted with permission from Ref. [2] Copyright 1986, American Chemical Society.

of CO_2 in the condensed phase after expansion to low pressure as a function of reduced pressure and reduced temperature of the supercritical fluid. The solid line gives the minimum reduced pressure and temperature required to avoid passage through a two-phase region for any stage of an isenthalpic process. The dashed lines give the indicated fractions of fluid as a condensed phase after expansion to atmospheric pressure. The reduced conditions necessary to avoid a two-phase system given in this figure also provide guidance for other one-component fluid systems. Recent laser light scattering studies [4] of supercritical fluid expansions have verified the predicted conditions shown in **Figure 2** for the formation of a two-phase product in supercritical CO_2 and N_2O. For more complex fluids with greater numbers of internal degrees of freedom, the maximum temperature at which a two-phase separation occurs in an isenthalpic expansion is decreased. The thermodynamic prediction of a two-phase product is also consistent with observations of extensive cluster formation in the electron impact ionization mass spectra of carbon dioxide at lower expansion temperature [5].

Approximate agreement of predictions from **Figure 2** with experiment is also demonstrated for the expansion of supercritical water shown in **Figure 3**. In this experiment, water at 600 bar was expanded through a stainless steel restrictor (1.0 mm long by 0.1 mm i. d.) at various fluid temperatures (T_e). The result is a gas or steam jet with an average droplet size that decreases as fluid temperature increases, causing the steam jet to become less visible. At the highest temperature shown (515 °C), just above the region where a two-phase product is anticipated ($<$ 510 °C), only the slightest hint of a visible jet exists even though the flow rate is about 20 ml/min (liquid).

10.2.1 Estimation of Flow Rates Through SFC Restrictors

In addition to providing analyte transfer to the detection region, a primary requirement for restrictors is to provide acceptable flow rates or SFC linear velocities over the range of pressures relevant to a given separation. For short restrictors, the flow rate during an adiabatic expansion cannot exceed the speed of sound until within a few nozzle diameters of the end of the capillary [6, 7]. A method of estimating flow rates of compressible fluids through straight tubes based upon friction losses has been described [2] and leads to calculations of a flow reduction coefficient, F_r [7], based upon the gas and the "equivalent capillary length". **Figure 4** gives values for F_r in terms of the capillary length to diameter ratio (L/D or aspect ratio) applicable to fluids such as CO_2 and H_2O. Also shown in **Figure 4** is the maximum capillary outlet/inlet pressure ratio (P_2/P_1) for sonic velocities at the capillary exit. When P_2/P_1 is below the minimum for sonic flow (as is usually the case), then the pressure at the capillary exit (P_{exit}) can be estimated because the

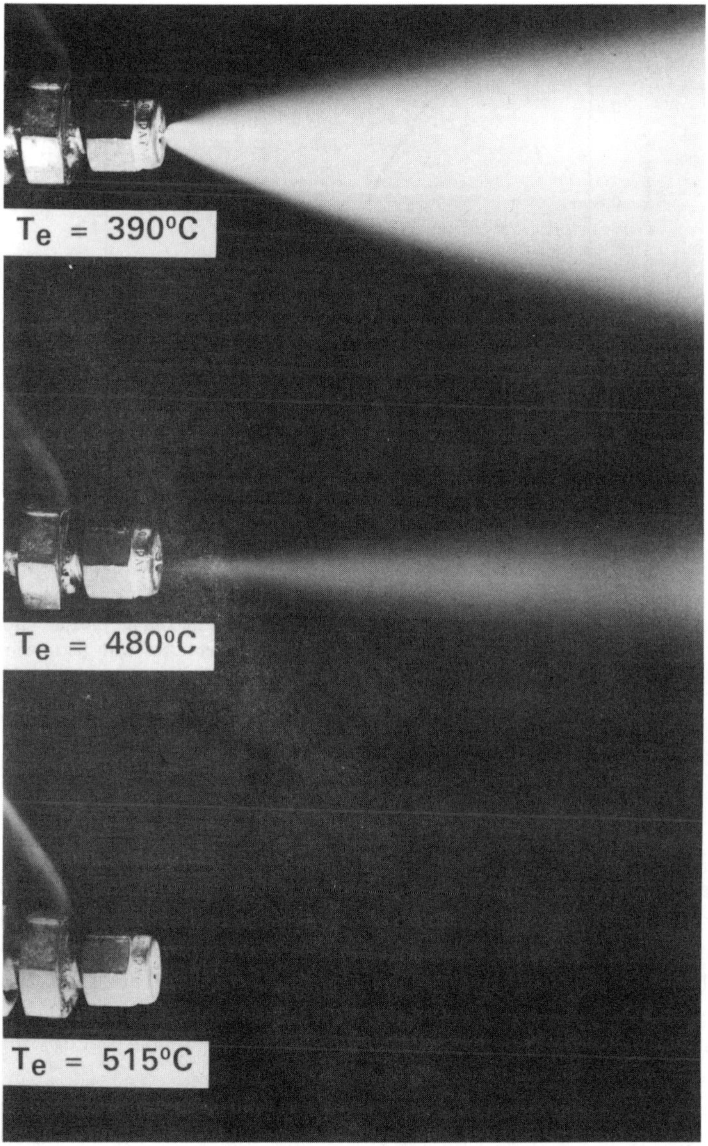

Figure 3
Photographs of water expansions from 600 bar at various pre-expansion temperatures (T_e) showing the liquid droplet size decrease rapidly as temperature is increased. Reprinted with permission from Ref. [2], Copyright 1986, American Chemical Society.

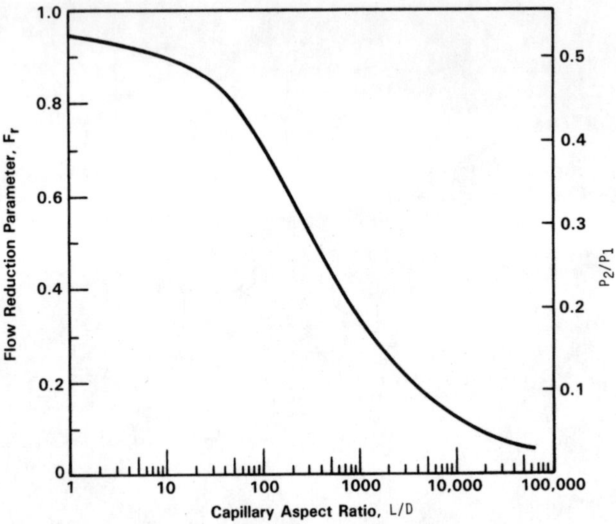

Figure 4
Flow reduction coefficient as a function of capillary aspect ratio (L/D) for an adiabatic expansion to sonic velocities at the capillary exit. Also shown is the maximum pressure drop ratio (P_2/P_1) resulting in sonic conditions at the capillary exit. Reprinted with permission from Ref. [2] Copyright 1986, American Chemical Society.

ratio becomes P_{exit}/P_1 [2]. Reasonable estimates of flow rates (±30%) were demonstrated using this method [2].

For adiabatic expansions the fluid temperature at the capillary exit can be calculated, which allows the fluid density at the capillary exit to be estimated [7]. The exit density can be surprisingly high. For example, a pentane mobile phase at 250 °C and 240 bar (ρ = 0.47 g/cm³) at a flow rate of 150 μl/min (liquid) that is expanded through a 6.4 mm x 9 μm i.d. restrictor has an exit density of approximately 0.12 g/cm³. Under these conditions the average fluid density (0.29 g/cm³) can be used to estimate the average velocity in the restrictor to be 8400 cm/s, corresponding to a residence time of 7.5 x 10^{-5} s. This approach allows the fluid solvating power at the capillary exit to be estimated.

A one-dimensional compressible flow model [8] has been used to predict an incomplete pressure drop over tapered and straight-walled restrictors when operated under choked flow (sonic exit velocities) conditions. The remaining decompression takes place outside the restrictor, which maximizes fluid density and minimizes analyte nucleation in the restrictor. The same model also predicts the pressure drop will be lower (which gives a higher exit fluid density) for a tapered restrictor than for a straight-walled restrictor giving the same mass flow rate at constant temperature and pressure.

When long restrictors (straight-walled) are used to obtain the slower flow rate needed for more optimal capillary SFC, the assumption of an adiabatic process becomes invalid due to the longer restrictor residence times. However, the speed of sound limitation to maximum velocity still applies. The analyte will precipitate and ultimately plug a long restrictor if it is not heated to vaporize the analyte or cooled to maintain sufficient density for solubility. Studies [2] have also shown that considerable variation in flow rates of CO_2 as a function of pressure can occur when significant deviations exist from the adiabatic assumption.

Increased restrictor temperature reduces the flow rate (mass flux) if the fluid is a gas. If the fluid is cooled to the liquid state in the restrictor, mass flux will increase with temperature. This behavior is consistent with the temperature dependence on fluid viscosity. Studies [2] have shown that at higher temperatures ($>$ 120 °C) the linear velocity of CO_2 has a nearly direct dependence on pressure. If the SFC column and restrictor are independently thermostated, SFC mass flux (linear velocity times column density) should be independent of column temperature. **Figure 5** shows the SFC mass flux at six pressures as a function of column temperature for a restrictor heater temperature of 275 °C. At lower pressures the expected behavior was observed with mass flux nearly independent of SFC temperature. At higher pressures an increasingly significant decrease in mass flux was observed at higher SFC temperatures. This result implies that the effective restrictor temperature was increased by the greater mass flux. Such behavior is consistent with data obtained at a higher restrictor heater temperature (400 °C), shown normalized to the lower temperature data in **Figure 5** (dashed lines). The lower mass flux and the higher restrictor heater temperature reduced the heating effect of the fluid. These results support the assumption that heating can be inefficient for short restrictors due to both the high mass flux and the poor thermal conductivity of silica.

The relationship between restrictor temperature and mass flux also explains previous observations of SFC linear velocity either increasing or decreasing during pressure programming. A constant temperature fluid entering the restrictor results in a linear dependence of mass flux on density. However, if the SFC temperature is close to the critical point, the density increase in the SFC column with pressure can be much greater than the increase in mass flux through the restrictor. Under such conditions the linear velocity decreases as pressure is increased. Thus, it is possible to select restrictor temperature (as far as compatible with detection requirements) to control variation in linear velocity during pressure programmed operation.

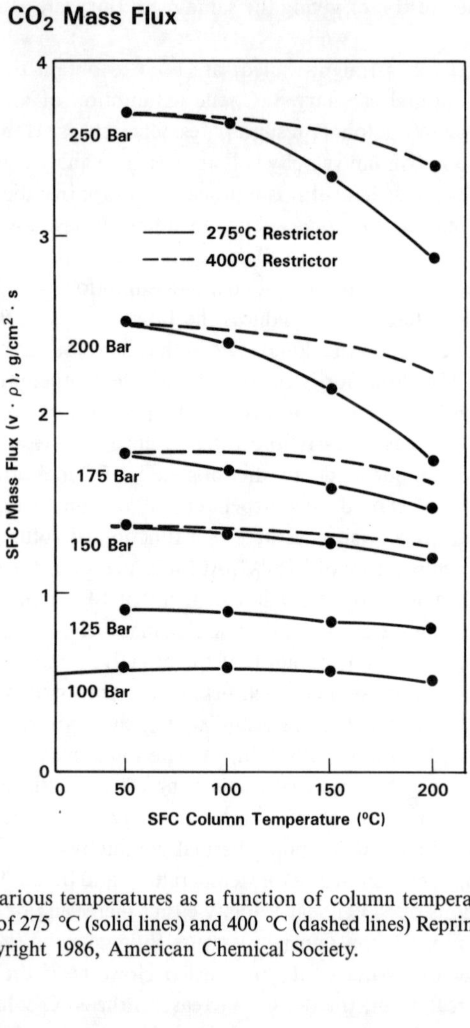

Figure 5
SFC mass flux at various temperatures as a function of column temperature for a restrictor heater temperature of 275 °C (solid lines) and 400 °C (dashed lines) Reprinted with permission from Ref. [2], Copyright 1986, American Chemical Society.

10.3 Transport of Nonvolatile Compounds Through Restrictors

A vital property of SFC restrictors is their effectiveness for transporting "non-volatile" compounds to various gas-phase detectors. Success depends upon the restrictor design, fluid pressure, and temperature, the particular demands imposed by the detection method (*e.g.,* flame ionization or chemical ionization mass spectrometry), and, under some conditions, the volatility or melting point of the compounds. It has been observed [9,10] that flame ionization detection (FID) of less-volatile or higher-molecular-weight compounds can cause large spikes in the signal with a frequency on the order of 0.1 – 10 s – 1. Similar phenomena can also be observed under some conditions (usually for a very hot ion source) with MS detection. It has been suggested [11] that the origin of this behavior is due to ''clustering'', and it was demonstrated [9] that the incorporation of a suitably long time constant (a few seconds) with FID can produce a reasonable chromatogram. It has also been observed [12] that increased fluid temperature delays the onset of the spiking phenomena to later eluting (and typically less-volatile) compounds, and suggested a ''spiking delivery rate threshold'' that varies with solute volatility. Heating of the fluid through relatively long restrictors ($L/D > 10^3$) with FID improves analyte detectability [10], as does heating a region just prior to the restrictor [13]. The performance of drawn restrictors is improved by heating of the fluid prior to the drawn restriction and perhaps by better heat transfer through the thinner walls of the tapered portion of the restrictor.

For truly nonvolatile compounds, two distinct modes of operation appear feasible, the first leading to the spiking by a ''precipitation'' process, and the second resulting in a ''nucleation regime'' where it is unlikely for FID. In the spiking mode, the nonvolatile analyte tends to collect on restrictor walls and, if the local temperature is above the analyte melting point, flows towards the end of the restrictor. The liquid will collect at the end of the restrictor and be periodically entrained in the high shear gas flow and transported to the detector. This is a complex process with details that likely depend on the liquid viscosity and restrictor geometry. Solid particle agglomerates also appear to form by a related mechanism. These processes often lead to plugging of the restrictor.

The fluid expansion process has been directly observed, and various materials expanded from supercritical fluid solutions through fused silica capillaries have been collected and analyzed. In one set of experiments polycarbosilane was studied [14]. The molten solute was observed to collect at the fused silica capillary exit and be periodically entrained in the gas flow [2]. For polycarbosilane in pentane at 350 °C and 100 bar, short fibers were formed (**Figure 6a**). The high shear forces can apparently also elongate polymer droplets to yield fibers. The particle size (either fibers or spheres – depending upon the solute, fluid, and restrictor conditions) was found to increase with capillary diameter. These observations are consistent with the observed FID spiking phenomena and suggest a process involving the restrictor walls and solute melting point. The average particle size formed with 5 – 10 μm restrictors (slightly larger than typical for capillary SFC) is in the 0.5 to 5 μm size range.

The second mode of operation producing much smaller particles occurs when analyte nucleation is delayed to near the end of the restrictor. **Figure 6b** shows

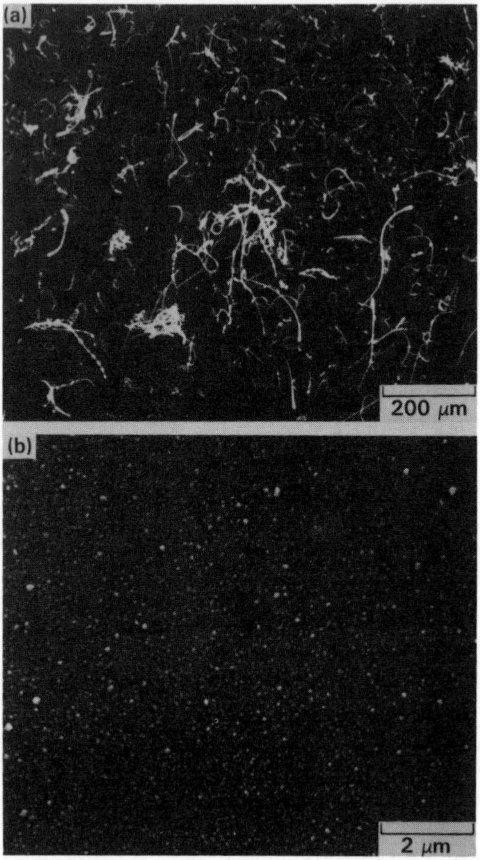

Figure 6
Photographs showing products obtained from the expansion of supercritical polycarbosilane-pentane solutions. (a) fibers collected from a solution expanded from 350 °C and 100 bar, and (b) ultrafine polycarbosilane powder formed by expansion from 250 °C and 240 bar.

polycarbosilane collected under such a set of conditions (240 bar and 250 °C). In this case, large quantities of particles with diameters of 0.02 to 0.04 μm were collected [2]. These small particles evidently form rapidly ($< 10^6$ s) and apparently pose no difficulty for FID. The particle sizes formed from concentrations more indicative of SFC for individual analytes at the restrictor (< 1 ppm), however, are too small to be resolved by conventional electron microscopy (*i. e.* < 200 Å diameter). The major requirement for operation in the "nucleation mode" is to maintain solvating conditions to near the end of the restrictor, which is facilitated by the use of short restrictors and fluid conditions enhancing solubility. It is important to recognize that this does not necessarily correspond to elevated restrictor temperatures.

Increased pressure always serves to enhance the transport of nonvolatile analytes, but in practice, pressure is limited by either the SFC conditions needed to yield sufficient retention or the maximum pressure for the system. The effect of temperature on fluid phase solubility is more difficult to predict, but some guidance is provided by analyte vapor pressures. Heating of the fluid just prior to expansion and heating of the restrictor itself were shown to improve and extend detection for compounds having some (low) volatility [2]. (An operational definition of volatility is the ability to thermally desorb an analyte from a solid inlet probe to obtain a mass spectrum). It was shown that volatility plays a major role in detection of such analytes, both directly and indirectly, because fluid phase solubilities (at a given density) generally mirror analyte vapor pressures [2,15]. Other workers have made similar observations for compounds of low volatility [8]. For FID detection of truly nonvolatile compounds, heating of the fluid before or during expansion can be counterproductive, because of lower fluid densities and the resulting lower fluid phase solubilities. Such behavior was observed for the analysis of polycarbosilane with the greatest detector response at the lowest detector temperature [2].

10.3.1 Cluster Formation During Expansion

Additional understanding of the expansion process, particularly as it applies to SFC-MS interfacing to a chemical ionization source, can be gained by consideration of cluster formation during expansion of a high-pressure jet through a nozzle. When an expansion occurs in a chamber with a finite background pressure, the expanding gas will interact with the background gas, producing a shock wave system. This includes barrel and reflected shock waves as well as a shock wave perpendicular to the jet axis (the Mach disk). The Mach disk serves to heat and break up small solvent and solute-solvent clusters formed during the expansion [5,16,17]. Recent laser studies have supported these observations by suggesting that solute-solvent clusters, which are readily detected in the absorption spectra, do not persist beyond the Mach disk [18]. The break up of clusters can be further aided by the rapid collisional processes in the mass spectrometer chemical ionization source, where the average molecule undergoes on the order of 10^5 collisions before ionization, assuming no wall losses.

The distance from the restrictor orifice to the Mach disk may be crudely estimated from experimental work [19] as 0.67 $D(P_1/P_2)$ 1/2, where D is the orifice diameter and P_1 is the fluid pressure and P_2 is the expansion region pressure. For a capillary restrictor the fluid pressure is more correctly described by the fluid pressure at the end of the capillary (P_{exit}), which can be estimated as previously described, and represents typically 5–20% of the SFC column pressure. Thus, if P_1 = 60 bar, P_2 = 1 torr, and D = 5 μm, the distance to the Mach disk is ≈ 0.7 mm.

Because initial cluster formation involves primarily volatile solvent molecules, heat applied in the later stage of this expansion reduces solvent clustering but does not necessarily facilitate transfer of solute molecules. As already demonstrated, it is clear that as the capillary restriction becomes longer, solute clustering during decompression can become significant, leading to precipitation of the solute. This can be manifested as "spikes" in the detector signal or plugging of the restrictor. Although nucleation processes will ultimately result in solute clusters, this problem can be overcome to some extent using sufficiently short restrictors and SFC analyte concentrations that avoid fluid conditions yielding a two-phase system upon expansion. It is even feasible that solute-solvent association during the later stage of the expansion may assist in delaying the solute nucleation process. The details of the expansion process remain unresolved, but their study appears vital to development of improved interfaces for nonvolatile solutes.

10.4 Practical Restrictors for SFC

Figure 7 schematically illustrates the wide range of restrictor geometries that have been investigated for SFC and SFC-MS. In addition to the straight-walled capillary restrictor (**Figure 7A**, tapered fused silica restrictors (**B**) are widely used and readily produced (although significant user skill is required). They are made by a controlled drawing of a heated region of fused silica tubing from which the polyimide coating (shown as dark lines in **Figure 7**) has been removed. Drawn capillaries or other variations (**C**) resulting in larger effective aspect ratios, typically in the range of 10^2–10^5, can be relatively easily fabricated and are less likely to plug than straight-walled capillaries [2,10]. The fluid expansion through such restrictors can be approximated with the relationships described earlier [2] by using an exit orifice diameter slightly larger than the actual diameter. A practical advantage of the drawn restrictor, compared to a similar length of small diameter capillary, results from the fact that the undrawn portion of the capillary allows effective heating of the fluid prior to expansion. This also eliminates the need for a zero-dead-volume union in a heated region. A major disadvantage of the drawn restrictor is its fragility, which results from removal of the polyimide coating.

Other restrictor designs yielding shorter restrictors with more favorable (i.e., smaller) aspect ratios and giving high exit fluid densities are shown in **Figure 7, C–F**. Fast drawing techniques have also been used with fused silica to produce restrictors with short, small diameter orifices (**C**). These restrictors can be fabricated

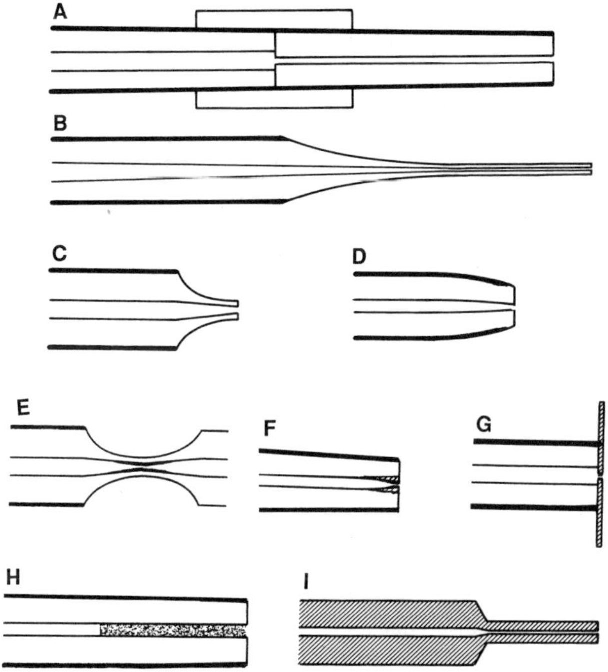

Figure 7
Cross-section diagram (not to scale) of various restrictor designs that have been explored for SFC. All restrictors except the pinhole laser-drilled orifice (G) and the pinched restrictor (I) are fabricated from fused silica tubing coated with polyimide (dark lines).

in a relatively reproducible fashion, but have the drawback of the drawn region being fragile. The polished "integral" restrictor (**D**) is fabricated by polishing the closed end of a piece of fused silica capillary (obtained by careful heating) until a small orifice giving the desired flow rate is obtained [20]. A similar variation, the converging-diverging restrictor, (**E**) is formed by compressing the column wall near the capillary exit at elevated temperatures and retaining a short length at normal diameter to focus the expanding jet [21]. Another approach (**F**) being investigated in several laboratories involves internal deposition of a material to largely close the end of a capillary. These techniques produce relatively robust restrictors; however, their production can be a somewhat tedious process.

The diaphragm, or laser drilled "pinhole" orifice (**G**), was explored in earlier SFC-MS work and has severe practical limitations for capillary interfaces due to the alignment and low dead volume connection requirements [16]. Properly used, this approach provides a nearly optimum expansion process, and is feasible with the higher flow rates of conventional diameter packed columns. The thin diaphragm, however, must be carefully supported to withstand high SFC pressures.

The porous frit restrictor (**H**) is quite different from the other restrictors [22]. The frit resembles a column packed with submicron particles, resulting in multiple fluid paths with aspect ratios $>> 10^5$. The frit restrictor is relatively rugged and will not be easily plugged by particles entrained in the fluid. Fluid velocity in the porous frit restrictor is always subsonic, and the tortuous flow path, combined with the relatively long residence time, provides for more efficient heat transfer to the expanding gas. This results in improved transport for sufficiently volatile compounds, but prevents detection of compounds with insufficient vapor pressures at the restrictor temperature. Thus, it is expected that truly nonvolatile materials will not be transmitted through this restrictor and will ultimately result in plugging. Given the drawbacks of other restrictors this does not constitute a major limitation for lower temperature (and less solvating) fluids.

The "pinched" restrictor (**I**) is fabricated by carefully crimping $\approx$ 1 mm length of the termination of a segment of capillary platinum-iridium tubing (50 to 100 μm i. d.) so as to obtain the desired flow rate [16]. This approach has the disadvantage of being irreproducible and giving unstable behavior, particularly at low flow rates. An alternative method of producing restrictors from platinum-iridium tubing that have dimensions similar to "**F**" was also described, but appears quite difficult [23]. The advantage associated with these restrictors is their compatibility with higher temperatures and more polar fluids (*e. g.*, supercritical water or ammonia) than fused silica restrictors.

Microscope photographs of two popular restrictor designs, the tapered and polished restrictor, are shown in **Figure 8**. These photographs illustrate the relative fragility of the tapered restrictor and relative robustness of the polished restrictor. The cone formed in the end of the polished restrictor, which allows the orifice diameter to be adjusted by careful polishing of the restrictor end, can be easily observed. Polishing of this restrictor tip has only proceeded to the point where the tip of the cone was just barely opened to yield a very small orifice.

10.5 Qualitative Restrictor Evaluation

A qualitative evaluation of the performance of various restrictor types can be achieved by comparing the chromatograms of a challenging sample (*e. g.*, high molecular weight and low volatility) obtained with the different restrictors using the same chromatographic apparatus and conditions. However, caution must be exercis-

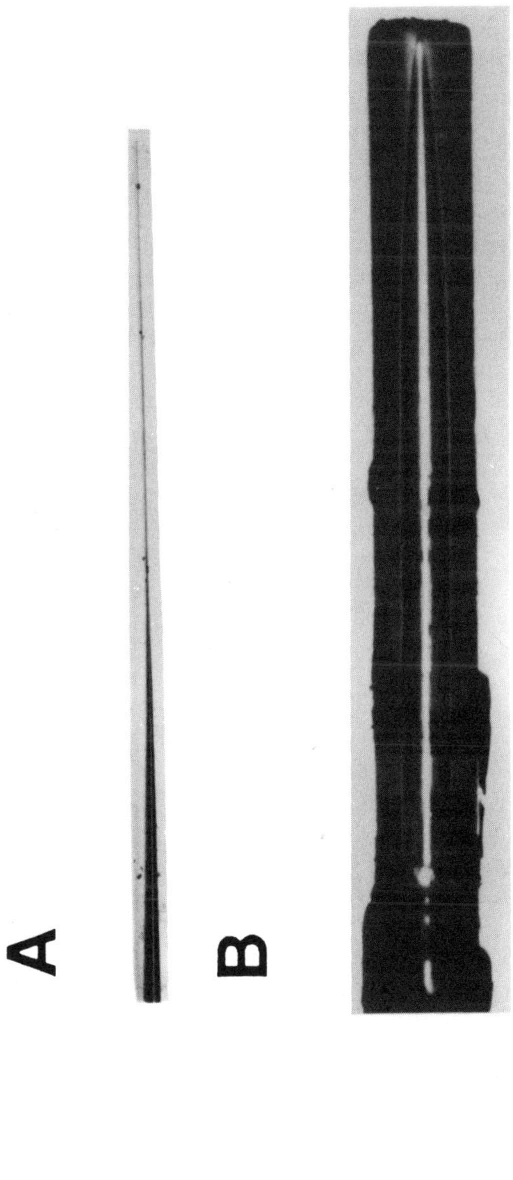

Figure 8
Microscope photographs of (A) a tapered restrictor (10) and (B) a polished restrictor (50) both drawn from 50 μm i.d. fused silica.

ed when comparing restrictor designs, because significant differences in the performance of apparently identical restrictors have been observed. This can be probably be accounted for by slight geometrical differences in the actual restrictor and the positioning of the restrictor in the detector region.

The chromatograms from the evaluation of four different restrictor types are shown in **Figure 9**. In this evaluation, a polyether polyol of relatively low volatility with an average molecular weight of approximately 2000 dalton was used as a test sample. Chromatographic separations were obtained using a 10-m x 50-μm-i. d. column coated with crosslinked SE-54 stationary phase and a pure carbon dioxide mobile phase at 125 °C. Where possible (tapered and polished), the restrictors were fashioned directly on the column end. Flame ionization detection was utilized and the detector base was heated to 400 °C. The linear velocity of the mobile phase was adjusted to be similar for each restrictor type. The ability of the various restrictors to transport increasingly higher molecular weight material can be easily observed by counting the individual oligomers that were eluted. However, since split injection was used for these analyses, possible injection discrimination effects could also influence the peak profiles.

A porous frit restrictor was used for the chromatogram shown in **Figure 9 A**. Only the first eleven oligomers were effectively transported through the restrictors as evidenced by the broadened and nonexistent peaks of the later eluting oligomers. This behavior clearly illustrates the dependency on analyte vapor pressure to transport materials through this type of restrictor. A straight-walled restrictor of 10-μm-i.d. was used for the chromatogram shown in **Figure 9 B**. Spiking problems can be observed as early as the fourth oligomer. In addition, the peaks were significantly broadened and exhibited poor symmetry after the onset of the spiking behavior. A higher concentration of the polymer (2 to 3 times) was injected for this separation compared to the other separations shown in **Figure 9**, as can be observed by the relative peak heights of the first eluting oligomers. Even with more material injected, the peaks for the later eluting oligomers disappeared rapidly illustrating the poor transport of the less volatile oligomers through the restrictor. A tapered restrictor, in fact the restrictor shown in **Figure 8 A**, was used for the chromatograms shown in **Figure 9 C** and **9 C-1**. Over twenty-five oligomers were eluted in the chromatogram shown in **Figure 9 C**. Additional oligomers might have eluted if the mobile phase solvating power had been further increased. However, 400 bar is the practical upper pressure limit. The last well-defined oligomer peak actually eluted after the pressure was at its upper limit. No visible signs of spiking or peak broadening were observed. The partial chromatogram shown in **Figure 9 C-1** was obtained with the same tapered restrictor. Spiking and decreased peak height (relative to the chromatogram in **Figure 9 C**) were observed beginning with approximately the sixteenth oligomer with complete disappearance of the peaks by the twenty-second oligomer. The poorer performance of the restrictor in this situation can probably

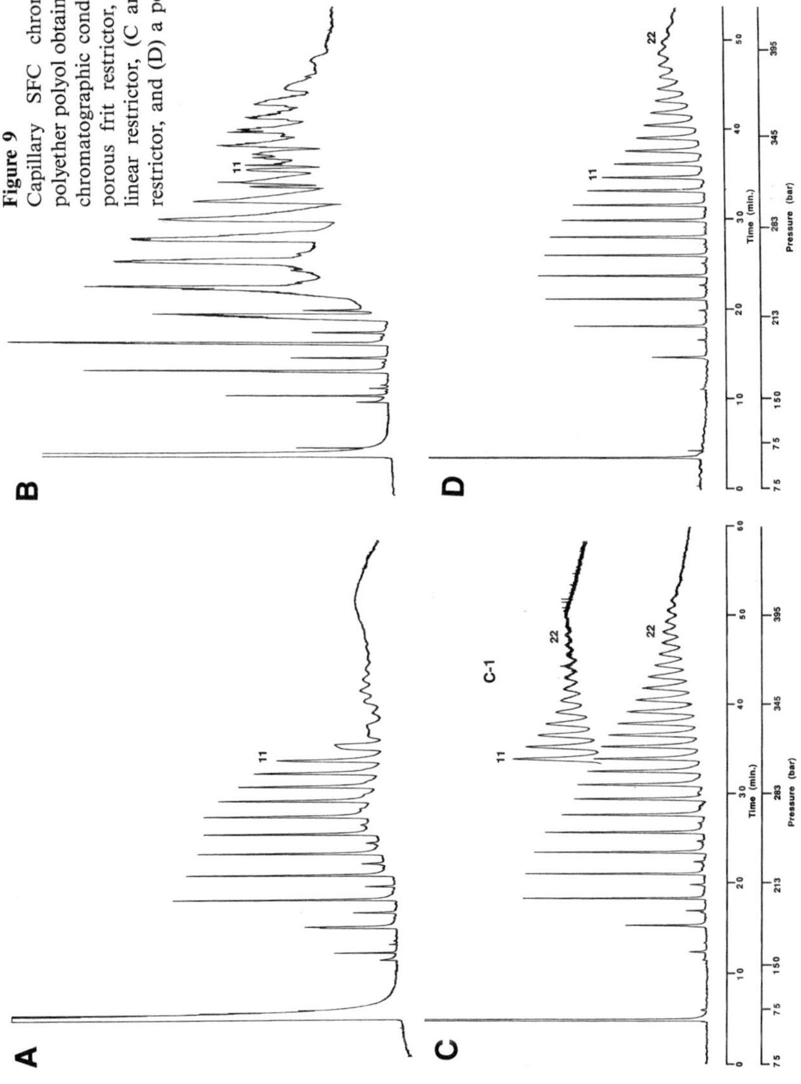

Figure 9
Capillary SFC chromatograms of a polyether polyol obtained under identical chromatographic conditions using (A) a porous frit restrictor, (B) a 10-μm-i.d. linear restrictor, (C and C-1) a tapered restrictor, and (D) a polished restrictor.

be accounted for by its position in the detector base. By slightly repositioning the restrictor, the greatly improved performance observed in the chromatogram shown in **Figure 9C** was obtained. Consequently, it can be concluded that minor differences can create significant changes in performance. A polished restrictor was used for the chromatogram shown in **Figure 9D**. Reasonable performance was also obtained for this restrictor with approximately twenty-three well-defined oligomer peaks observable. However, loss of peak height and degradation of the peak shape was noticeable by the twenty-first oligomer.

The tapered restrictor clearly provided the best performance for transport of the higher-molecular-weight and less volatile oligomers to the detector. The polished restrictor also provided reasonable performance. It is interesting to note that the restrictor behavior also affected the peak shape and baseline characteristics of nearly the entire chromatogram rather than just the final end of the chromatogram.

10.6 Conclusions

The performance of the restrictor and the efficiency for transfer of less volatile analytes to the gas phase during expansion for subsequent detection is crucial to SFC and SFC-MS operation. The success of transporting analytes to the detector depends on the volatilities and fluid phase solubilities of the analytes, the restrictor design, and fluid pressure and temperature. **Table 1** summarizes observations relevant to the various restrictor designs and provides an estimate of their effective aspect ratios. Clearly, the ideal restrictor has not yet been developed and, due to its important role in SFC detection, remains an area of active research. The major consideration in selection of a restrictor should be the volatility of the analytes; less volatile analytes are expected to benefit from a restrictor with a low aspect ratio and operation which maximizes fluid density at the restrictor exit. Operation at higher flow rates can also provide an alternative approach for improving detection of less volatile analytes. For analytes volatile at restrictor temperatures, the most rugged and conveniently replaceable restrictor is desired.

Table 1
SFC restrictors summary.

	Type [a)]	Effective Aspect Ratio [b)]	Fabric-ation	Reproduc-ibility	Primary Advantage	Primary Disadvantage
A.	Straight Walled Capillary	≈ 2000 – 20,000	–	Excellent	Easily assembled	Requires union to capillary, excessive precipitation of non-volatiles
B.	Long Drawn Capillary	100 – 10,000	Easy	Good	Easily fabricated, moderate aspect ratio	Fragile
C.	Fast (Short) Drawn	50 – 1000	Easy	Good	Low aspect ratio	Fragile
D.	Polished	100 – 2000	Moderate	Fair-Good	Low-moderate aspect ratio, rugged	Possible plugging of narrow orifice
E.	Converging-Diverging	100 – 2000	?	?	Low-moderate aspect ratio, rugged	Possible plugging of diverging opening
F.	Internal Deposition	100 – 2000	Tedious	?	Low-moderate aspect ratio, rugged	Difficulty in fabrication
G.	Pinhole (laser drilled)	5 – 25	Difficult	Poor-Fair	Near ideal expansion	Difficult alignment and elimination of dead volume
H.	Porous (multipath) Frit	$>> 10^5$	Moderate	Good	Commercially available rugged	Loss of nonvolatiles
I.	Pinched	?	Easy	Poor	Applicable with nearly any fluid	Unstable and irreproducible

(a) All fabricated from fused silica tubing except for the pinhole orifice substrate (G) and pinched platinum iridium restrictor (I).
(b) Appropriate range for comparison with straight walled capillary (A).

Acknowledgments

This work was supported by the U.S. Department of Energy, Offices of Fossil Energy and Health and Environmental Research, under Contract DE-AC06-76RLO 1830. Pacific Northwest Laboratory is operated for the U.S. Department of Energy by Battelle Memorial Institute. The authors thank *A. J. Kopriva* for technical contributions to this work.

References

[1] *A. B. Cambel* and *B. H. Jennings,* "Gas Dynamics", McGraw Hill, New York (1958).
[2] *R. D. Smith, J. L. Fulton, R. C. Petersen, A. J. Kopriva,* and *B. W. Wright*, Anal Chem. **58** (1986) 2057–2064.
[3] *R. D. Smith* and *H. R. Udseth,* Anal. Chem. **55** (1983) 2266–2272.

[4] *S. R. Goates, N. A. Zabriskie, J. R. Simons,* and *B. Khoobehi,* Anal. Chem. **59** (1987) 2927–2930.
[5] *R. D. Smith, H. R. Udseth,* and *H. T. Kalinoski,* Anal. Chem. **56** (1984) 2971–2973.
[6] *H. W. Liepman* and *A. Roshko,* " Elements of Gas Dynamics", Wiley, New York (1957).
[7] *C. E. Lapple,* Trans. Sm. Inst. Chem. **39** (1943) 385–432.
[8] *R. W. Bally* and *C. A. Cramers,* HRC & CC **9** (1986) 626–632.
[9] *J. C. Fjeldsted, R. C. Kong,* and *M. L. Lee,* J. Chromatogr. **279** (1983) 449–455.
[10] *T. L. Chester, D. P. Innis,* and *G. D. Owens,* Anal. Chem. **57** (1985) 2243–2247.
[11] *J. C. Giddings, M. N. Meyers, L. McLaren,* and *R. A. Keller,* **162** (1968) 67–73.
[12] *T. L. Chester,* J. Chromatogr. **299** (1984) 424–431,
[13] *B. E. Richter* HRC & CC **8** (1985) 297–300.
[14] *S. Yajima, Y. Hagegawa, J. Hayashi,* and *M. Iimura,* J. Mater. Sci. **13** (1978) 2569.
[15] *R. D. Smith, H. R. Udseth,* and *B. W. Wright,* in *J. M. L. Penninger, M. Radosz, M. A. McHugh,* and *V. J. Krukonis* (eds) "Supercritical Fluid Technology", Elsevier, Amsterdam (1985) pp. 191–223.
[16] *R. D. Smith, J. C. Fjeldsted,* and *M. L. Lee,* J. Chromatogr. **247** (1982) 241–243.
[17] *R. D. Smith, W. D. Felix, J. C. Fjeldsted,* and *M. L. Lee,* Anal. Chem. **54** (1982) 1883–1885.
[18] *S. R. Goates, A. J. Barker, H. S. Zakharia, B. Khoobehi,* and *C. W. Sheen,* Appl. Spectrosc. **41** (1987) 1392–1397.
[19] *R. D. Smith, H. T. Kalinoski, H. R. Udseth,* and *B. W. Wright,* Anal. Chem. **56** (1984) 2476–2480.
[20] *E. J. Guthrie* and *H. E. Schwartz,* J. Chromatogr. Sci **24** (1986) 236–241.
[21] *C. M. White, D. R. Gere, D. Boyer, F. Pacholec, and L. K. Wong,* HRC & CC **11** (1988) 94–98
[22] *K. E. Markides, S. M. Fields,* and *M. L. Lee,* J. Chromatogr. Sci. **24** (1986) 254–257.
[23] *K. Grob,* HRC & CC **6** (1983) 178–184.

11 Industrial Significance of Supercritical Fluid Chromatography

H. Michael Widmer,
Department of Analytical Research, Ciba-Geigy Limited, CH-40002 Basel, Switzerland

Summary:

For many years our analytical research group has been involved in the industrial application and promotion of supercritical fluid chromatography. Since 1982 we have been building our own equipment and developing the appropriate instrumentation for density gradient work. As commercial hard and software increasingly becomes available, SFC will be introduced at the application front, and we predict a major breakthrough in its industrial exploitation.

In the light of our experience within a wide spectrum of chemical compounds and problems, ranging from pharmaceutical to agrochemical products and from dye chemicals to plastics and additives, we are now witnessing a period of expansion for SFC technologies. Depending on the analyst's requirements, both capillary as well as packed column technology will contribute significantly to the state of the art.

The enthusiasm with which we promote capillary as well as microbore column SFC within our company is the consequence of our vast experience in industrial problem solving. We believe that there are four areas in which SFC may establish itself as an industrial stronghold:

- **Capillary SFC challenges the latest developments in high-temperature capillary gas chromatography as shown herein by comparison of chromatograms obtained by both techniques.**
- **SFC assumes a superior position to LC whenever solvent elimination becomes an issue, including coupling techniques such as SFC/FT-IR, SFC/MS, and SFC/TEA and supersonic jet molelcular beam experiments.**
- **SFC may be combined with supercritical fluid extraction techniques, thus offering appropriate on-line combinations, eliminating tedious sample preparation procedures.**
- **According to our own experience, certain problem compounds exhibit a superior solubility in supercritical fluids as compared with the traditional solvents used in liquid chromatography.**

The versatility and significance of SFC are demonstrated in a number of exemplary industrial investigations.

Concentrate:

This chapter documents some uses of SFC for industrial problem solving. Analytical results obtained by both SFC and high temperature open-tubular GC on the same samples are compared.

11.1 Introduction

Our group has been involved in **supercritical fluid chromatography (SFC)** [1–5] since 1982 and the appropriate experience has been accumulated through active research and promotion of SFC as well as through analyzing typical products of our company. It should be mentioned that others have developed the fundamental aspects of the technique while our group has been primarily involved in developing applications of SFC to the solution of industrial problems, and, to a certain extent, instrument development. Our firm is active worldwide in the field of pharmaceutical and agrochemical products and is a producer of dye chemicals, plastics, and additives. For a numer of years we built our own SFC equipment, but during the last 1 1/2 years we started to use the available commercial hard and software.

It might be of interest to know that chromatography predominates all the other techniques in our analytical laboratories, occupying more than 50% of the analytical activities, including those in routine laboratories. About one third of all our compounds may be conveniently investigated by gas chromatography (GC), but the majority are either polar or represent high molecular weight compounds and therefore are not suited for GC; they must be analyzed by high performance liquid chromatography (HPLC) or alternatives thereof.

From the very beginning we supported packed column SFC as well as capillary techniques (CSFC). However, the predominant portion of our packed column methodology was focused on microbore column technologies, which were mainly used in coupled systems with mass spectrometry (SFC/MS) [5, 6], Fourier transform infrared spectroscopy (SFC/FT-IR) [7, 8] and the Thermal Energy Analyzer (SFC/TEA) [9–11]. There is an extensive sharing of know-how with another research group within our department devoted to the development of microbore column HPLC/MS [12]. This group is presently most successfully involved in industrial problem solving through LC/MS.

Within our department we support several research programs, covering chromatography as well as alternative separation techniques. This background together with the vast application front of our company puts us in an excellent position to evaluate SFC and to compare it with the existing alternatives, such as capillary gas chromatography (CGC) and HPLC.

Our enthusiasm about SFC may be taken as a consequence of our experience. We consider the potential of SFC to be more than a chromatographic niche in the multifaceted field of industrial analytical chemistry. It offers a welcome addition to CGC and HPLC.

In industry we should not argue about different technologies. Economic considerations and technical advantages offer a clear-cut decision about the question of which analytical technique should be used to solve a particular analytical problem. Economic and technical superiority are the driving forces for methodological replacement. There are no excuses to stick with a well established but obsolete method since an instrument generation lasts only for about 3 to 4 years, and it is always an economic decision to replace inferior chromatography equipment after 7 to 8 years of use. Therefore it takes about an equally long period to introduce a new methodological technology, which can reach its full potential only after the commercial hard- and software is available for routine applications [4].

In our view there are mainly four fields in which SFC will establish an industrial stronghold:

- SFC is a true challenge for the latest developments in **High-Temperature Capillary GC** (HTCGC).
- SFC holds a superior position whenever **Solvent Elimination** is a key issue, including condensed-phase chromatography coupled with MS, FT-IR, TEA [9–11] and similar detection system techniques.
- SFC offers a technological follow-up of **Supercritical Fluid Extraction** (SFE), which may be significant in modern sample preparation.
- According to our experience, a significant number of compounds exhibit problematical **solubilities** in most solvents used in HPLC, but may be dissolved in supercritical fluids.

Our optimism about SFC persuades us that it will grow into a standard analytical technique of comparable size to present-day GC.

11.2 Supercritical Fluids in the Context of the Opportunities and Needs of Modern Analytical Chemistry

For about three decades, analytical chemistry has been in a phase of dynamic transition. Its opportunities and therefore its practical significance have grown enormously, mainly due to the fast advances in analytical instrumentation. The limits of detection as well as the time required to perform an analytical investigation have been reduced by orders of magnitudes and instrumental precision has replaced the manual skills of the analyst. Through instrumental ease and simplification, the analytical work is shifting from the specialist laboratory to the spot where the problems arise. Real-time investigations become reality and they allow diagnostic statements about a chemical system, thereby challenging the more traditional quality control measurements that dominated in industry only a few years ago, but were limited to retrospective findings. However, the modern technologies, characterized by frequent repetitive measurements (surveillance monitoring) provide the basis for diagnostic and prognostic statements.

This modern approach of analytical chemistry enhances the possibility of integrating analytical chemistry in to process control. However, the growing potential is accompanied by needs for improvements in the state of the art. The appetite of the industrial manager for more and better analytical performance is awakened and the analysts are challenged to continue their research efforts in this direction [13, 14].

The increased pressure for improved analytical performance also exposes the bottleneck of the present state of the art. However, analytical chemistry has always been a science that has proved its flexibility and adaptability to real-world problems. Once more it will live up to the present and future demands.

In order to fully evaluate the significance of SFC we must review it in the light of its contributions to satisfying modern demands, *i.e.* the future state of the art in analytical chemistry. This procedure focuses realistically on the strengths and potential weaknesses of SFC.

Table 1 summarizes the essential features and problems of the present and future state of the art in analytical chemistry and **Table 2** outlines the bottleneck that must be overcome in the near future. We try to include the contributions SFC can make as well as those of its potential challengers.

The present state is dominated by tedious sample preparation techniques, followed by chromatographic separation steps before quantification. Furthermore, most investigations are performed in a batch mode and stepwise approach.

The future state of the art will be dominated by an integration of sample preparation into an automatic on-line procedure, *i. e.* flow systems. The batch mode will be replaced by modern high frequency surveillance technologies. The stepwise approach will be replaced by coupled systems and ultimately integrated total analysis systems. Therefore, chromatographic procedures will be challenged by flow injection or at least combined with them.

11.3 Capillary SFC

The strength of CSFC [15] lies in its ability to resolve the components of many complex samples, and its ability to elute nonvolatile and/or thermally labile analytes. In its performance it resembles CGC [16–18], but in no way is it limited

Table 1
Present and future state of the art in analytical chemistry

Present state	Future state	Contributions
Scientific aspects		
Sample preparation	Integrated sample preparation	**SFC, FIA:** Simplification, time savings
GC, LC, TLC	GC, LC, TLC, and SFC	**SFC:** More flexibility
Conventional spectroscopy	Molecular beam spectroscopy	**SFC:** Simplification in solvent elimination
Technical aspects		
Batch investigation	Analytical surveillance technology	**SFC, FIA:** Advantage over LC
Single step investigation	Coupled systems	**SFC:** Ease of coupling and interfacing
Applications		
Quality control	Quality steering Process steering	**SFC, FIA:** Advantage over LC; **Sensors**
Investigation in analytical lab	At the spot investigation	**SFC, FIA:** Advantage over LC; **Sensors**

FIA: Flow injection analysis.

to the separation of volatile components. The compounds typically investigated by CSFC are those without chromophores, but with reactive, temperature labile groups. CSFC handles substances of relatively high molecular weight and polymeric nature (oligomers), which often exhibit a poor solubility in the conventional solvents used in HPLC. However, in order to take full advantage of the available detecting system, CSFC is predominantly run without or only with small amounts of modifiers [19].

Today, sophisticated immobilization methods are used to prepare CGC columns. The same techniques fulfill the demanding needs for CSFC, although CSFC columns have smaller inner diameters, usually in the range of 50 to 100 μm [20]. To use these efficiently, density programs are a prerequisite and the appropriate hard- and software must be available for routine investigations [21].

The following examples illustrate the power and advantage of CSFC separations. First we present comparisons of results obtained by HTCGC [22, 23] and CSFC, the later being performed with polyethylene glycol (PEG) and SE-54 coated capillary columns, respectively.

Figure 1 represents the investigation of aorta descendens. It clearly demonstrates the superiority of CSFC over HTCGC, the later being characterized by column bleeding.

Table 2
The bottleneck of the present state of the art in analytical chemistry

Problems	Contribution from SF/SFC
Sampling/Sample preparation	
Mass transport	Efficient mass exchange between phases because of favorable diffusion in supercritical fluids
Solubility problems	
New solution media	Density may be enhanced due to the compressibility of supercritical fluids, increasing the solubility, replacing the need for HTCGC (density gradient technique)
Separation science	
General features	SFC combines excellent choice of various detector systems with possibility for chromatographing polar and high molecular weight compounds and to take advantage of specific separation effects of the mobile phase (*e.g.* ion pairing). This is particularly true when using carbon dioxide as the mobile phase.
Trace investigations	
complex matrices	Supercritical fluids are superior solvents, reducing the problem of solvent elimination; they enable an easy identification through SFC/MS, SFC/FT-IR, SFC/ICP, molecular beam technology, *etc.*

ICP: Inductively coupled plasma.
GPC: Gel permeation chromatography

Figures 2 and **3** refer to the HTCGC and CSFC investigations on a detergent mixture. The results indicate that CSFC provides as much and possibly more information than HTCGC. This impression is enhanced by **Figure 4,** where the chromatograms of polyethylene monolaurate obtained by HTCGC and CSFC are compared. The resolution of CSFC allows a sophisticated interpretation of compound groups as seen in **Figure 5 A** and **5 B**.

Figures 6 and **7** allow comparison of HTCGC and CSFC on surfactants. The same conclusions as above are reached concerning the information content of the chromatograms.

We conclude our comparisons between CGC and CSFC with a study on isocyanates. **Figure 8** shows an investigation with CGC, whereas **Figure 9** refers to the results obtained with CSFC. With CGC only a limited number of constituents are recorded, whereas with CSFC we recognize the efficient separation of an entire series of additional polymeric compounds, which are not eluted by CGC.

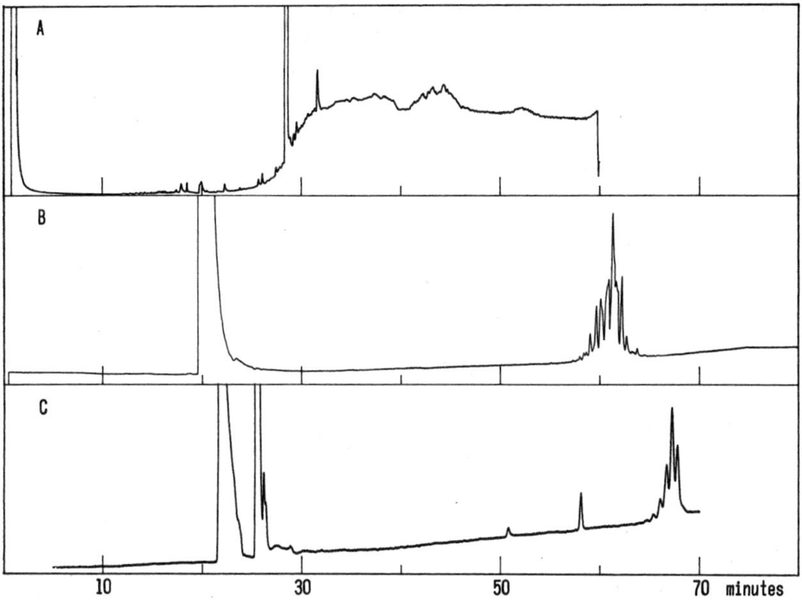

Figure 1
Chromatograms of aorta descendens:
A: HTCGC; B: CSFC: PEG stationary phase; CO_2; 100 °C; C: CSFC: SE-54 stationary phase; CO_2; 100 °C.

11.4 SCF with Packed Columns

The strengths of packed column SFC are manifold, including the relative large sample capacity, which makes it suitable for coupling with sophisticated structure elucidating techniques. Another characteristic is the possiblility to cover, in a single run and in a minimum of time, a range of substances with extremly different properties. These advantages are the result of the possibility to run density programs [21] and additionally influence the retention time by appropriate modifiers [24]. We demonstrate this point with the separation of antioxidants, represented in **Figure 10**. The mixture contains high molecular weight nonpolar components as well as ionic species (calcium salts). To fully separate all components of this mixture and to obtain quantitative results with HPLC at least two different chromatographic runs are necessary, because the polarity range of the substances involved is too large to handle it with a single column. To analyze this mixture with CSFC a chromatographic run of about 40 min is required and the calcium salt is not eluted at all

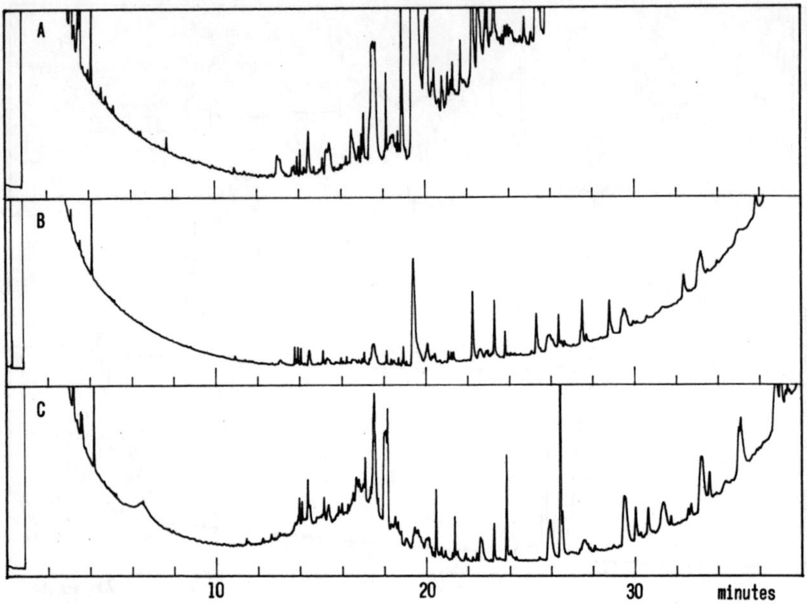

Figure 2
HTCG chromatograms:
A: Polyoxyethylene sorbitane trioleate; B: Polyethylene sorbitane monooleate; C: Polyoxyethylene sorbitane monopalmitate.

(**Figure 11**). However, the mixture is easily separated with packed column SFC in about 6 min, when a high modifier content is involved.

The analytical investigation of pharmaceutical and cosmetic ointments is often affected by a complicated sample preparation procedure, usually involving solvent extraction processes. We have demonstrated that such formulations are homogeneously dissolved in supercritical fluids such as CO_2 and therefore are easily analyzed by a combination of SFE and SFC. An example is given in **Figure 12**.

11.5 Hyphenated SFC Techniques

One of the most important advantages of SFC is its ease of coupling with sophisticated detection systems such as mass spectrometry and infrared spectroscopy.

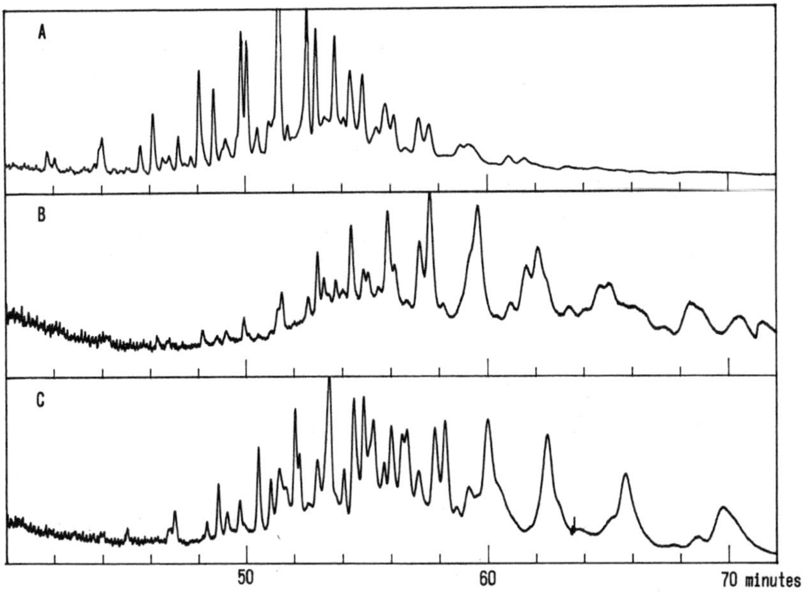

Figure 3
CSF chromatograms: PEG stationary phase; CO_2; 180 °C: A: Polyoxyethylene sorbitane trioleate; B: Polyoxyethylene sorbitane monooleate; C: Polyoxyethylene sorbitane monopalmitate.

A number of results from SFC/MS investigations are given in **Figures 13, 14, 15,** and **16**, covering a variety of different chemical compounds such as Irganox 1098 (MW = 636); Irganox 1010 (MW = 1176), Hygroton (MW = 339), and Serpasil (MW = 608).

Acknowledgment

The author gratefully acknowledges the experimental work performed by *Mr. Karl Grolimund* (HTCGC and CSFC work), *Dr. Steven Fields* (isocyanates) and *Drs. Peter Dätwyler, Aldo Giorgetti, Klaus Anton*, and *Mr. Nico Pericles* (packed column SFC, SFC/MS, and SFC/FT-IR results). Without their contributions this paper could not have been realized.

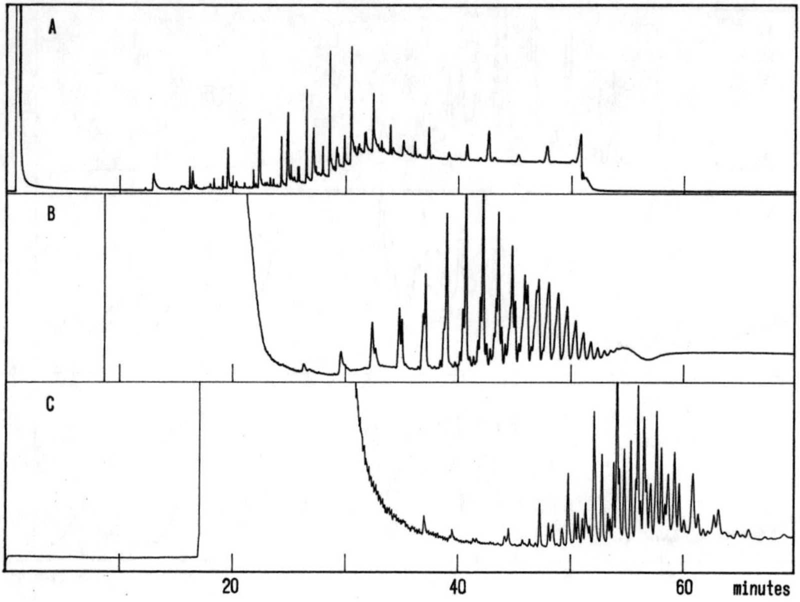

Figure 4
Chromatograms of polyethylene glycol monolaurate:
A: HTCGC; B: CSFC: SE-54 stationary phase; CO_2; 150 °C; C: CSFC: PEG stationary phase; CO_2; 150 °C.

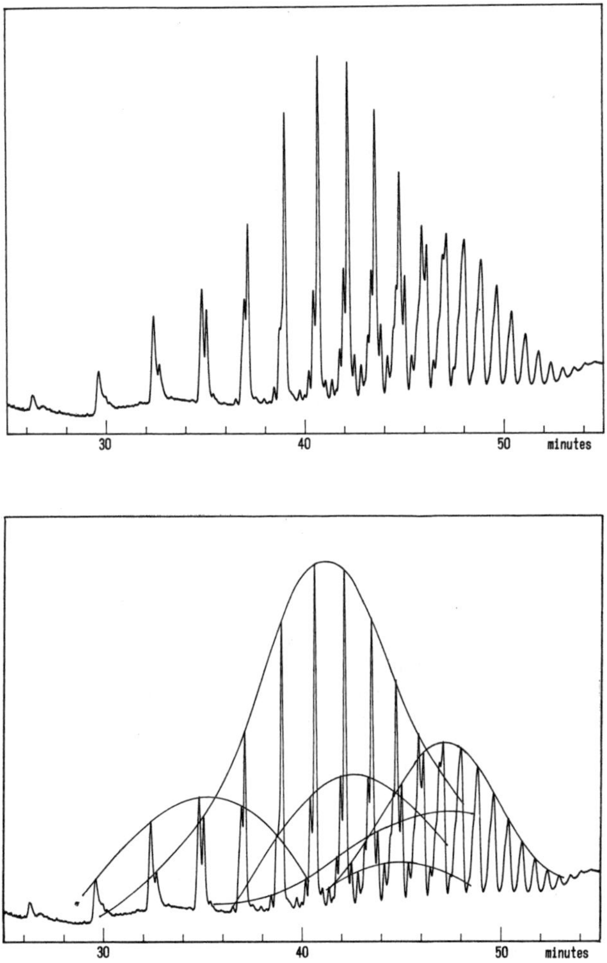

Figure 5 A and 5 B
SF chromatograms of polyethylene glycol monolaurate:
A: straightforward chromatogram; B: same chromatogram interpreted for compound families.

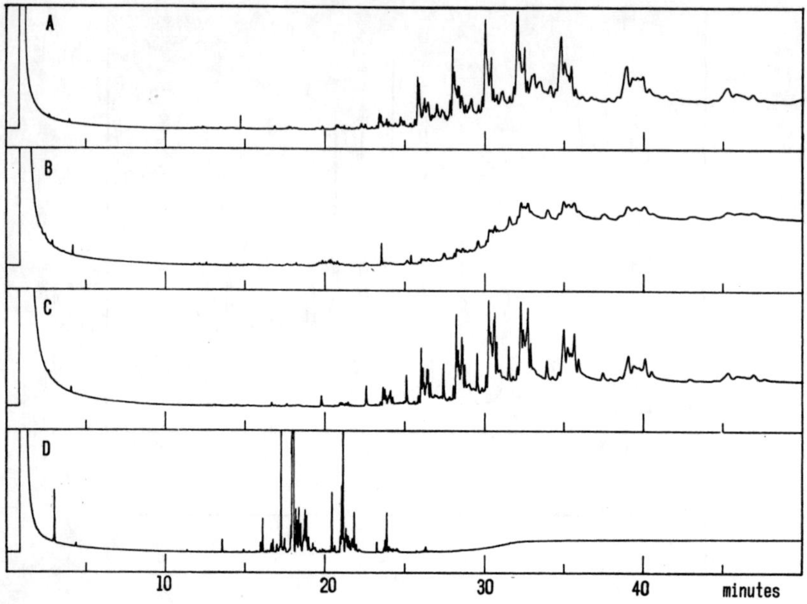

Figure 6
HTCG chromatograms of surfactants:
A: Surfactant mixture; B: Nonylphenol ethyleneoxide 15; C: Nonylphenol ethyleneoxide 8; D: Nonylphenol Ethyleneoxide 1.5.

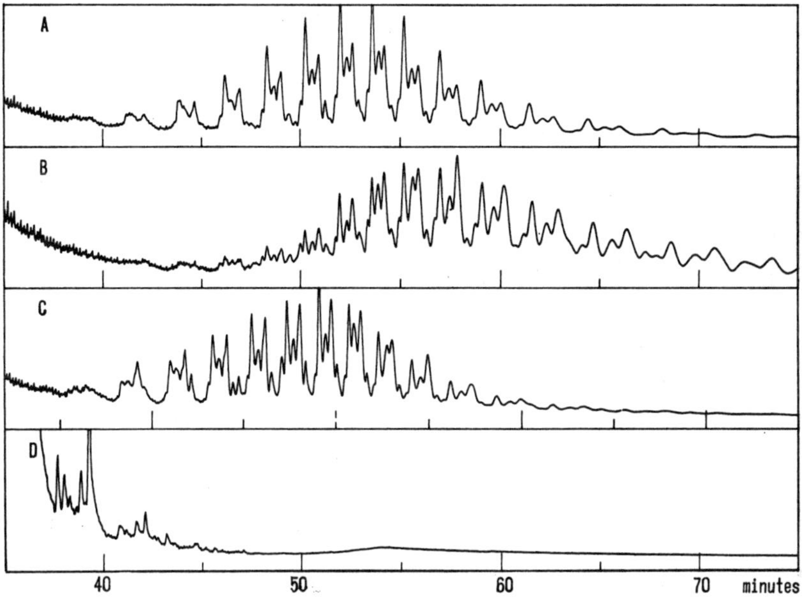

Figure 7
CSF chromatograms of surfactants; PEG stationary phase; CO_2; 180 °C
A: Surfactant mixture; B: Nonylphenol ethyleneoxide 15; C: Nonylphenol ethyleneoxide 8; D: Nonylphenol ethyleneoxide 1.5.

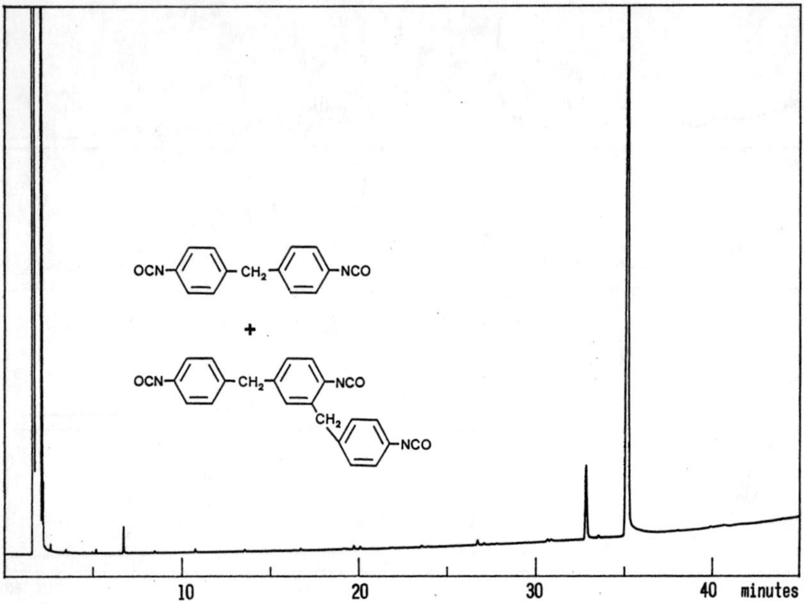

Figure 8
CG chromatogram of isocyanates.

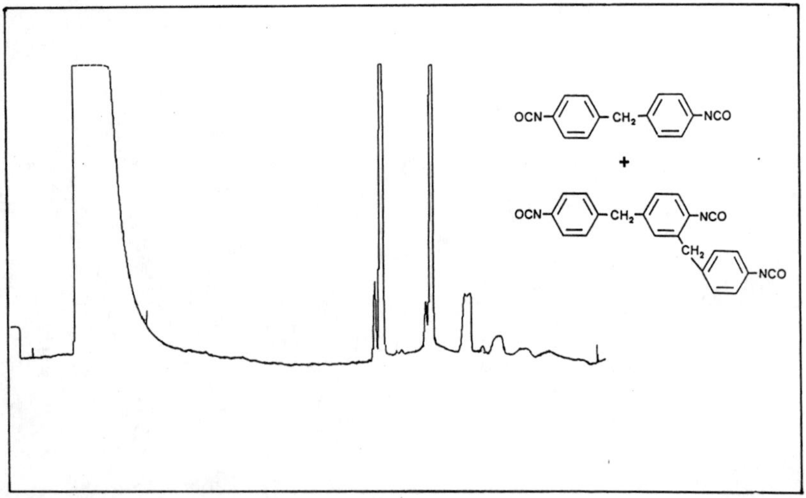

Figure 9
CSF chromatogram of isocyanates.

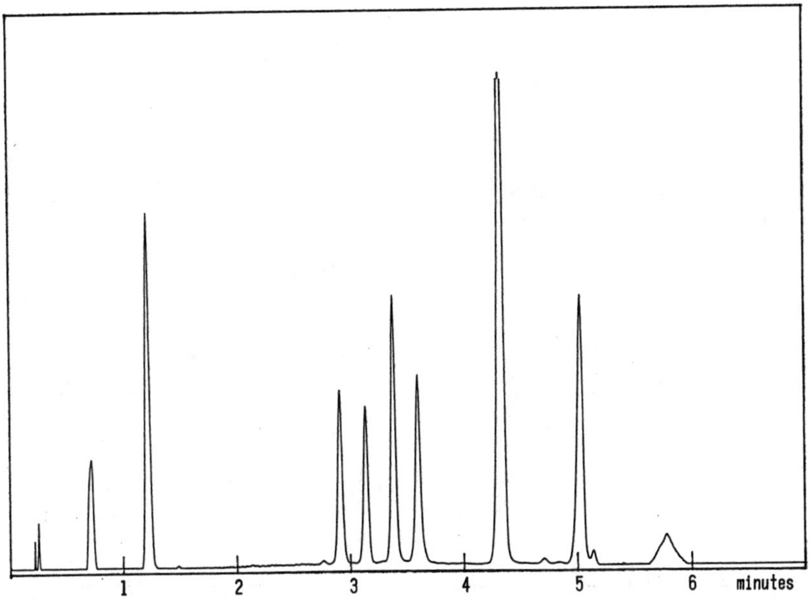

Figure 10
SFC of antioxidants with packed column and high modifier content.

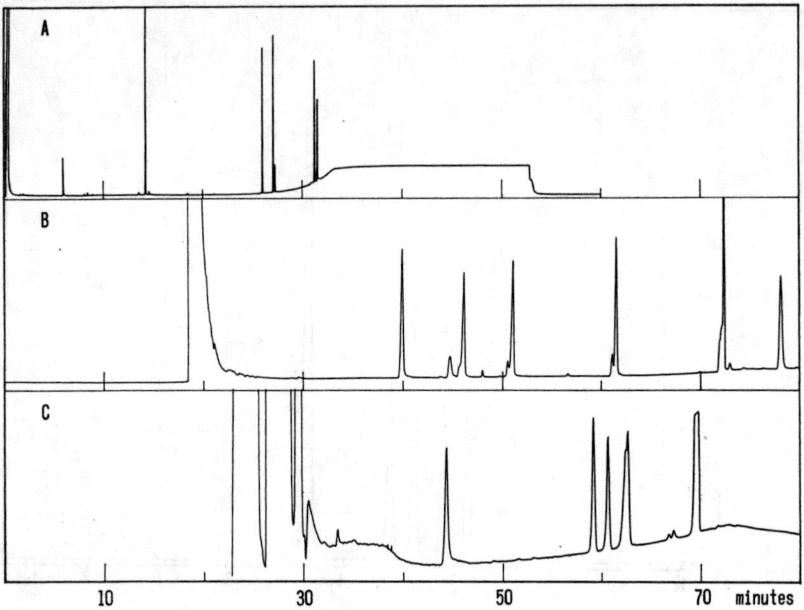

Figure 11
HTCGC and CSFC of antioxidants.

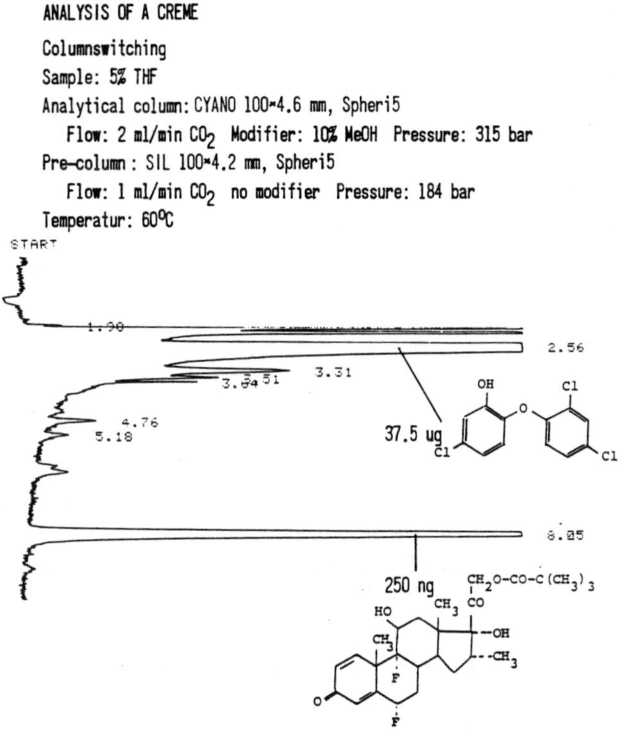

Figure 12
SFC of a pharmaceutical ointment.

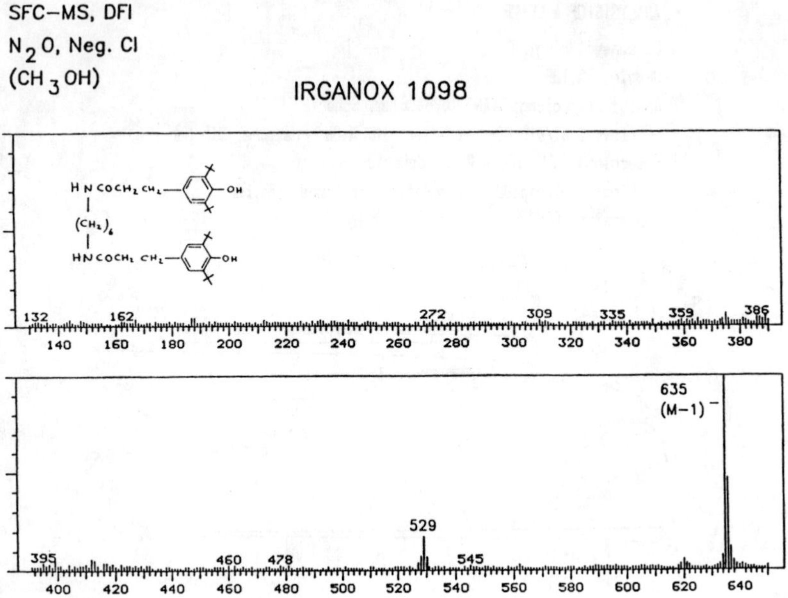

Figure 13
SFC/MS of IRGANOX 1098 by direct fluid injection; N_2O; negative chemical ionisation with CH_3OH.

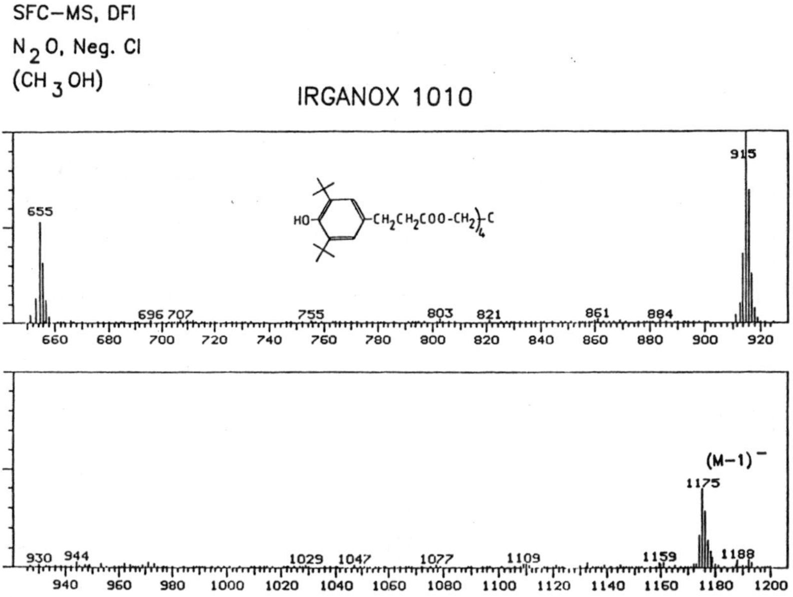

Figure 14
SFC/MS of IRGANOX 1010 by direct fluid injection; N_2O; negative chemical ionisation with CH_3OH.

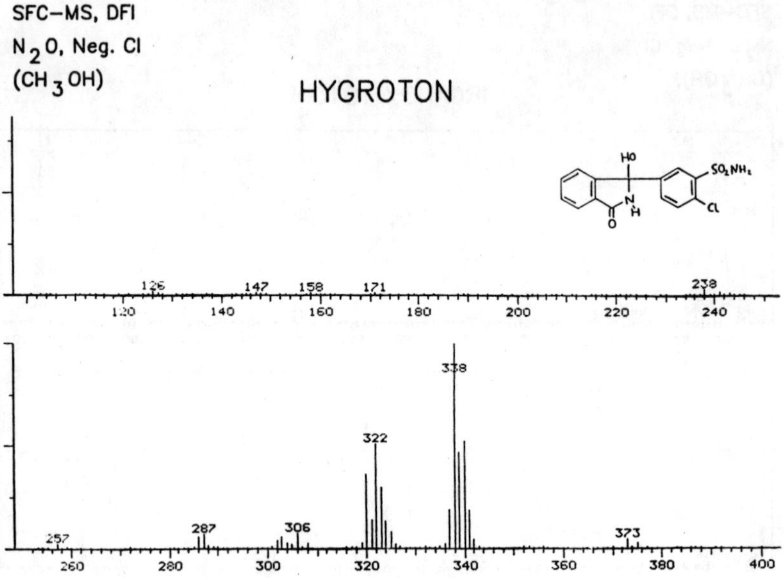

Figure 15
SFC/MS of HYGROTON by direct fluid injection; N_2O; negative chemical ionization with CH_3OH.

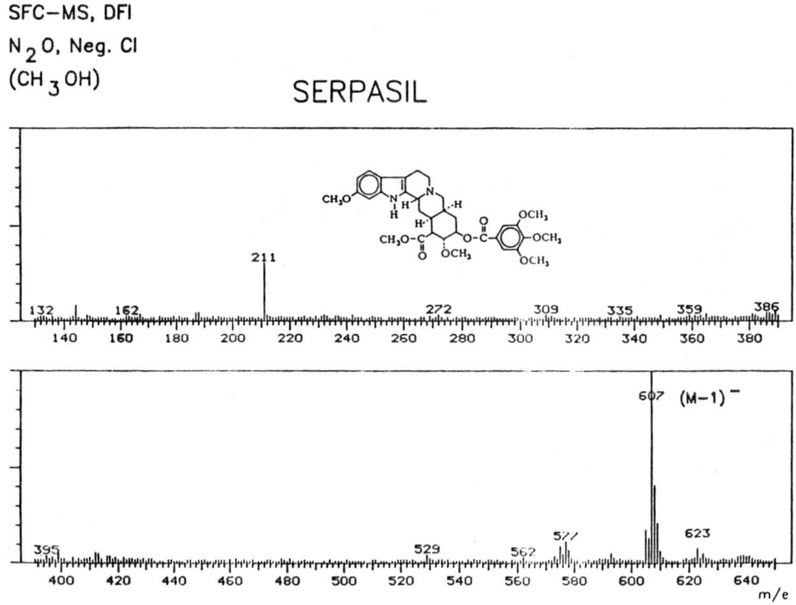

Figure 16
SFC/MS of SERPASIL by direct fluid injection; N_2O; negative chemical ionization with CH_3OH.

References

[1] *H. M. Widmer* and *P. Dätwyler*, Swiss Chem. **7** (1985) 16.

[2] *K. Grolimund, W. P. Jackson, M. Joppich, W. Nussbaum, K. Anton,* and *H. M. Widmer,* Proc. 7th Intern. Symp. Cap. Chrom. in Gifu, Japan, p. 625, (Es. *D. Ishi, K. Jinno,* and *P. Sandra*), University of Nagoya Press, Nagoya, Japan (1986).

[3] *H. M. Widmer* Chimia **41** (1987) 156.

[4] *H. M. Widmer,* Chimia **41** (1987) 391.

[5] *P. Dätwyler, H. J. Walther,* and *P. Hirter,* 4th Symp. on LC/MS, MS/MS and SFC/MS in Montreux, Switzerland, "A Capillary Interface for SFC/MS and DFI", October 1986.

[6] *R. D. Smith, W. D. Felix, J. C. Fjeldsted,* and *M. L. Lee,* Anal. Chem. **54** (1982) 1883.

[7] *A. Giorgetti* and *P. Dätwyler,* unpublished work.

[8] *P. R. Griffiths, J. A. de Haseth,* and *L. V. Azarraga,* Anal. Chem. **85** (1983) 1361 A.

[9] *H. M. Widmer* and *K. Grolimund*, Chimia **35** (1981) 117.

[10] *K. Grolimund* and *H. M. Widmer*, "Environmental Carcinogens" "Selected Methods of Analysis", p. 373, (Ed. *H. Egan*) IARC Sci. Publ. No. 45, Vol 6 "*N*-Nitroso Compounds"; Intern. Agency for Research on Cancer, Lyon (1983).

[11] *P. Hirter, H. J. Walther,* and *P. Dätwyler,* Poster presented at Symposium on Coupled HPLC, Saarbrücken (1985).
[12] *P. Hirter, H. J. Walther,* and *P. Dätwyler,* J. Chrom. **323** (1985) 89.
[13] *H. M. Widmer,* TRAC **2** (1983) VIII.
[14] *H. M. Widmer,* TRAC **3** (1984) VI.
[15] *K. E. Markides* and *M. L. Lee,* Advances in Capillary Chromatography (ed. J. G. Nikelly), Huethig, Heidelberg, Basel, New York (1986), 19.
[16] *H. M. Widmer* and *K. Grolimund,* ''Advances in Capillary Chromatography'', p. 95, (Ed. *J. G. Nikelly*) Dr. Alfred Huethig Verlag Heidelberg (1986).
[17] *E. Klesper, A. H. Corwin,* and *D. A. Turner,* J. Org. Chem. **27** (1962) 700–701.
[18] *M. Novotny, S. R. Springston, P. A. Peaden, J. C. Fjeldsted,* and *M. L. Lee,* Anal. Chem. **53** (1981) 407 A–414 A.
[19] *W. P. Jackson,* Ph. D. Dissertation, Brigham Young University, 1985.
[20] *S. M. Fields, R. C. Kong, J. C. Fjeldsted, M. L. Lee,* and *P. A. Peaden,* HRC & CC **7** (1984) 312.
[21] *J. C. Fjeldsted, W. P. Jackson, P. A. Peaden,* and *M. L. Lee,* J. Chromatogr. **289** (1984) 105.
[22] Prepared by *K. Grolimund.*
[23] Prepared by *K. Grolimund.*
[24] *K. Anton, R. Menes,* and *P. Dätwyler,* submitted for publication.

Subject Index